Elements of Robotics

Elements of Robotics

Greg Scott

STATES
ACADEMIC PRESS
www.statesacademicpress.com

States Academic Press,
109 South 5th Street,
Brooklyn, NY 11249, USA

ISBN: 978-1-63989-175-7

Cataloging-in-Publication Data

Elements of robotics / Greg Scott.
p. cm.
Includes bibliographical references and index.
ISBN 978-1-63989-175-7
1. Robotics. 2. Robots. 3. Automation. I. Scott, Greg.
TJ211 .E44 2022
629.892--dc23

For information on all States Academic Press publications
visit our website at www.statesacademicpress.com

STATES
ACADEMIC PRESS

Contents

Preface

Robotics is a branch of engineering that is related to creating fully or partially automated systems which can be used to replace humans. Primarily, robots are used in such environments where human life is subjected to the risk of death such as bomb deactivation and nuclear reactors. Robotics is the intersection of various fields such as electronics engineering, mechanical engineering, computer science, physics, etc. There are various types of robots such as domestic robots, medical robots, industrial robots, military robots and autonomous drones. All robots comprise basic units like actuators, powers supply, sensors and manipulators. This book unfolds the innovative aspects of robotics which will be crucial for the holistic understanding of the subject matter. While understanding the long-term perspectives of the topics, the book makes an effort in highlighting their impact as a modern tool for the growth of the discipline. Those with an interest in the field of robotics would find this book helpful.

Given below is the chapter wise description of the book:

Chapter 1- The branch of science which deals with the design, construction, operation and use of robots is known as robotics. There are several branches of robotics such as biomimetic robotics, swarm robotics, aerial robotics, ant robotics, cloud robotics, anthrobotics, cognitive robotics and evolutionary robotics. This chapter serves as an introduction to robotics and its various branches.

Chapter 2- The classification of robots can be done on the basis of various criteria such as the power source, workspace geometry, degrees of freedom, their application and environment. The topics elaborated in this chapter will help in gaining a better perspective about the classification of robots.

Chapter 3- A robotic design involves the creation of a plan or convention for the construction of a robotic system. Science and technology have undergone rapid developments in the past decade which has resulted in the discovery of significant tools and techniques in the field of robotic design. A few of such tools and techniques are physical design, design software, tools and equipment etc. which have been extensively detailed in this chapter.

Chapter 4- Robots are made up of many components. The key components of robots are actuator, end-effector, manipulator, servomechanism, stepper motor, and robotic sensors etc. This chapter has been carefully written to provide an easy understanding of the various components of the robots.

Chapter 5- Automation involves the use of a wide range of technologies and computer software to perform tasks that are usually done by humans. Vehicles that are capable of sensing their environment and moving safely without human intervention are known as autonomous vehicles. This chapter discusses in detail the theories and concepts related to robotics and automation.

Chapter 6- Bionics is the branch of science which applies the biological methods and systems found in nature to the study and design of engineering systems and modern technology. One such application of robotics is the shadow hand which is a humanoid robot hand system and is comparable to a human hand in size and shape. The diverse applications of robotics in the current scenario have been thoroughly discussed in this chapter.

Chapter 7- Degrees of freedom of a mechanical system refers to the number of independent variables required to define the position of a rigid body in space. The ethical problems that occur with robots fall under roboethics. All these diverse principles and laws of robotics have been carefully analyzed in this chapter.

Indeed, my job was extremely crucial and challenging as I had to ensure that every chapter is informative and structured in a student-friendly manner. I am thankful for the support provided by my family and colleagues during the completion of this book.

Greg Scott

1

Understanding Robotics

The branch of science which deals with the design, construction, operation and use of robots is known as robotics. There are several branches of robotics such as biomimetic robotics, swarm robotics, aerial robotics, ant robotics, cloud robotics, anthrobotics, cognitive robotics and evolutionary robotics. This chapter serves as an introduction to robotics and its various branches.

Robotics is an area where a number of scientific fields meet, and this fact already is a source of attraction for the involved scientists, for users, and the public. Expectations run high and in diverse directions. The word "robot" itself comes from literature and was created in the twenties by Czech poet, Karel Capek, in one of his plays, a play that ended tragically. In the forties, another writer, Isaac Asimov, made robots the leading figures in his utopian novels. Since these times, robots have been subjects of imagination. The reality of industrial robots only came in the sixties when Joseph F. Engelberger introduced the PUMA robot as a freely programmable, universal, handling device. With it came automation in manufacturing industry, economic issues, and social concern about human labour replaced by machines. The versatility of these robot machines has been increasing, largely due to their continuously increasing ability of information processing. The ultimate goal was the autonomous robot. However, as the application field for robots is widening, and the robot is coming out of the factory halls, new challenges are seen, and even a change of paradigm is taking shape. The robot is expected to be an extended, intelligent tool for the human, it should become a partner instead of being a "competitor" in fulfilling tasks, and there is a developing relation to biological systems. This development is illustrated by terms such as behaviour, emotions, or intelligence, taken out of their biological context and used to describe technical features and properties. For example, the term "intelligent" is being used to describe advanced robot behaviour, maybe still rather as a marketing term, but the idea certainly is to give it more meaning.

It is obvious that there are high expectations as to the future potential of robotics, even euphoric ones and somewhat unrealistically utopic. On the other side, there are sceptical views, seeing robotics as one of the most powerful technologies of the 21st century, together with genetic engineering and nanotech, threatening to make humans an endangered species. A more moderate and realistic, but still fascinating approach has been taken by a study group, consisting of experts from engineering, medical, philosophical and legal sciences, discussing the provoking question whether humans could be substituted by robots.

As robots of the future are supposed to show some kind of intelligence let us first look at various meanings of that term. A history of intelligence names a number of different approaches to explain and to categorise intelligence and its aspects.

In biology, intelligence is considered to support survival and to further natural selection. It thus, in principle, could be extended from humans and mammals to bacteria or plants as well. In physics and chemistry about anything can be retraced to the skillful arrangement of molecules, and certainly the research on the signal transmission between neurons in a brain has been profiting from this attitude.

In communication theory there are measures for intelligence based on entropy-related values. In social sciences the term "collective intelligence" characterizes the performance of knowledge handling structures such as libraries, legal systems, databases, or even the complex behaviour of an ant population. Artificial Intelligence, a spring-off from computer science, as it stands now, appears to require some embodiment to make intelligence actually work. The anthropocentric approach puts the human into the centre of consideration and regards intelligence as a capability of humans, as the core of their social competence, and as the ability to deal with reasons. Obviously, there is no way to come to a generally accepted definition of intelligence. Probably there is no need either, as there are other terms, such as "information", which has never been defined unanimously, and nevertheless it is the basis for a flourishing science.

The term robot intelligence could be justified by its usefulness to characterize some very desirable, human-like features, to be implemented in robots. In particular, a robot that should be useful to humans as a tool or even a partner, should be able to communicate with its human users on a reasonably high level, it should be able to "understand". Certainly, there are various levels of understanding that may have to be defined, giving measures of such intelligence. This stepwise approach may be less stringent than the classical definition of artificial intelligence using the Turing test but it could be more pragmatic and constructive. Thus, the ability for communication with humans could be the basis of anthropocentrically oriented robot intelligence.

Trends and Expected Benefits

The leading role of robotics is based on its inherent technology potential and, in particular, its relations to areas beyond technology. In comparison, the direct economic impact of robotics appears to be rather small. As robotics is a multidisciplinary area, expectations are very diverse as well.

Technology

Robotics can be regarded as a typical and representative part of Mechatronics, as a cutting edge technology in this rapidly expanding research field. Mechatronics combines in a synergetic way the classical engineering disciplines mechanical and electrical engineering and computer science, leading to new kinds of products. It can be stated that any technical progress in robotics will quickly spread over to products of everyday life and may eventually initiate further progress.

Automotive technology for modern cars, for example, in making advanced use of sensors for controlling their dynamics and assisting in safe driving are following ideas from robotics. In addition to that, the need for low-priced sensors in mass-produced cars has subsequently spurred the industrialization of micro technology in a very sustainable way.

Methods of robotics and mechatronics serve, beyond the individual product, as guidelines for the development of complete systems. Thus, the name system robotics or embedded robotics has been coined, to describe the integration of sensors, control, actuators and information processing into a system. This can be a car, an automated traffic control system, a military air defence system, medical service and human care systems, or the safety and energy management system of a building. There are already names such as cartronics, or domotronics, characterizing these new fields.

A very promising area is nano-techniques. Results from physics research are already available, but exploiting and using them on an industrial scale needs highly automated processes, it needs the transfer of technology known from robotics. In addition, this technology will be the basis for novel products in medical techniques, for techno-implants, or for prostheses.

An actual research topic in robotics is the development of "soft computing", i.e. learning algorithms and the interpretation of uncertain data from unstructured environments with methods such as fuzzy logic, neural nets or genetic algorithms. The spread-over to smart machine technology, with self-calibration, self-diagnostics, and self-tuning control loops can already be seen. This will lead to improved safety, reliability, and maintenance procedures for such smart machines, and there the expected economic benefits are obvious.

Another important area that is profiting from the advances in robotics is the control of complex dynamical systems. Examples are humanoid robots, as well as vehicles, construction machinery, machine tools, or prostheses for limbs and hands. On one side, it is the non-linear, model based, adaptive control that makes novel machine tools with parallel kinematic structures feasible, together with hard real-time operating systems, being used already in mobile robots. On the other side, bio-inspired behavioural control will lead to intelligent mobile robots moving smoothly in unstructured environments. Ideas for such a kind of control architecture are derived from motor control in animals. The relation between robotics and biology, however, goes beyond that and will be considered subsequently.

Biology and Neuro-Science

Robotics has a very stimulating co-operation of mutual interest with biological information processing and neuroscience. On one side data processing in biological systems can play a role model for advanced robot systems, and on the other side, the proven methods and tools for modelling in robotics may help to better interpret and understand the rather descriptive input/output models usually being available in biology. Investigations on the walking of stick insects, the navigation of desert ants, the swimming motion of fish, and the flight techniques of bees are being related to mobile robots. This effort may be stimulated by the underlying assumption that the ability to master motions is an evolutionary component of intelligence.

Neuro-science, in particular when related to motor control, is using robotics to verify its models for walking or grasping motions. This will have implications for the better understanding of human motor control and the design of prostheses, technical replacements of sensory organs (techno-implants).

The speculative question arises - will humans eventually be replaced by robots at all? The subsequent consideration might contribute to an answer: Contrary to the passionately discussed potential of controlling human development by genetic engineering, the evolutionary change of humans and human life due to technical developments takes place in a dark spot of our realization. Our ability for information processing, moving around and conquering distance and remoteness, underwater and in space and overcoming biological deficiencies has dramatically changed just in the past one hundred years. Actually, it will be complemented now by gradually integrating intelligent robots as tools into our life, and by enhancing our life system by "spare part medicine", making use of techno-implants, i.e. artificial sensory organs, technical heart assist systems and prostheses. This development and its consequences are hardly reflected consciously. It can be assumed that humans will make the relevant decisions step by step within an evolutionary process.

Man-Machine Relations

Obviously, the relation of man to machine will be the key issue in dealing with intelligent machines. It has become a most complex area with technical, psychological, and socio-economic aspects, and its implications are leading to a change of paradigm in the objectives of robotics and to a new definition of a robot.

Communication and Emotional Behaviour

The interaction between human and computer is seen today as one of the topics in computer science, and these approaches certainly will form an essential part of the communication methods between human and robot as well. In addition to that, safety aspects will be of much more importance, as a misunderstanding or a mistake in the communication can have most serious consequences. Furthermore, the activity and mobility that can be exerted by a robot will allow a wider range of communication modalities. The robot can turn its attention actively to points of interest, it can explore strange situations, and it can actually "bring" information or objects. The observer, seeing a real, moving robot will get different impressions than just by looking at some animated simulation from virtual reality. A nice example is the toy dog from Sony. It is even supposed to express "emotions", for example by wagging its tail. Emotions may play an important role for man-machine communication, expressing expectations, summarising rational thinking, condensing information and representing it in an easily understandable way. This desire to generate human-like communications may be an argument for building humanoid robots whose body motion could carry "emotional qualities" that might be more easily interpreted by humans.

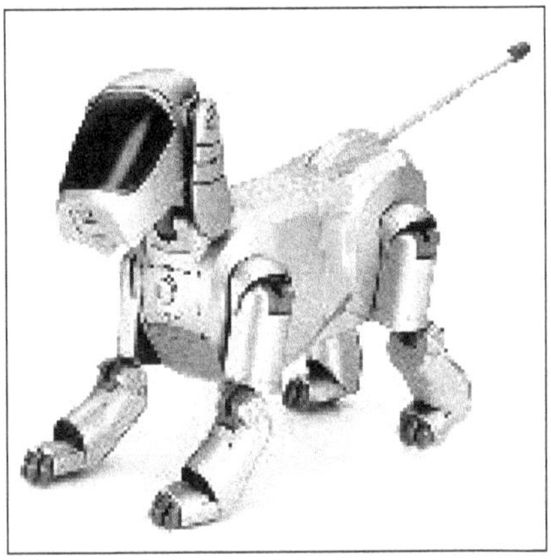

Aibo, Sony's robo-dog with emotional behaviours.

Allocation of Work and Authority

Another issue in designing the cooperation between man and machine is the allocation of work and the authority to make decisions when man and machine are jointly solving a task. Our capabilities to use automated machinery for carrying out complex tasks is increasing, and at the same time, we are aware of the limitations in the autonomy with which machines can or should perform these tasks. It therefore appears natural to design machines that can co-operate in an "intelligent" way with their human users, thus extending the range of the human and making best use of the capabilities of the machine. Such human oriented machinery will have novel features in their behaviour, related to their interaction with humans. The allocation of work between human and machine is a problem that is being discussed and will influence the way of automation and the design of future machines. We will need machines which can work in an autonomous way up to a certain degree of complexity, and in critical situations or on a higher level of autonomy the necessary interactions with the human operator or user have to be facilitated and structured.

The design of a machine and its motion control heavily depends on the degree of automation, i.e. how function and authority is allocated to man and machine. There are a number of criteria for allocation, briefly characterised, for example, by the terms left over, economic, comparison, and complementary. In the left over scenario, the degree of automation is as high as is technically feasible and only the tasks that cannot be automated are left over for the humans. Following only economic criteria means that for each partial task the cheapest solution will be chosen. In the case of comparison, a task will be allocated to the one, machine or man, who performs it best. The complementary approach assumes that man and machine have basically different capabilities that complement each other. The idea is to combine positive capabilities, the versatility of man and the consistency of the machine.

The robot system for charging the press brake has to fulfill the requirements of the MTO-concept. It has to offer user friendly and application-tuned task level programming for interactive program optimization and error correction, as well as interaction facilities to intervene in decisions on every system level. Thus, for example, the cell can be used for manual manufacturing of single, complex products during daytime, and for fully automated operation during night shift. In this way the expert knowledge and the responsibility of the press brake operator can be maintained and even enhanced. The human worker and the robot are sharing the same work space. This is different from classical industrial applications where, for safety reasons, the robot work space is separated from the human work space, usually by some kind of fence. An idealistic view for that holistic automation approach is shown in figure.

Still utopic scenario of a robot helper assisting a human expert in sheet bending.

Not only can the contents of work be shared between man and machine, the authority, as well. The different levels for the complementary allocation of authority, i.e. who is in control of a process, are shown in figure. The range spans from full human control (manual mode of operation) to the fully automated process. The examples mentioned in this figure are self-explaining, and additional examples from robot research have been shown in Schweitzer. Of course, safety is a dominant issue in designing the appropriate man-machine-interaction. It can be improved by making best use of the “intelligence” of the machine.

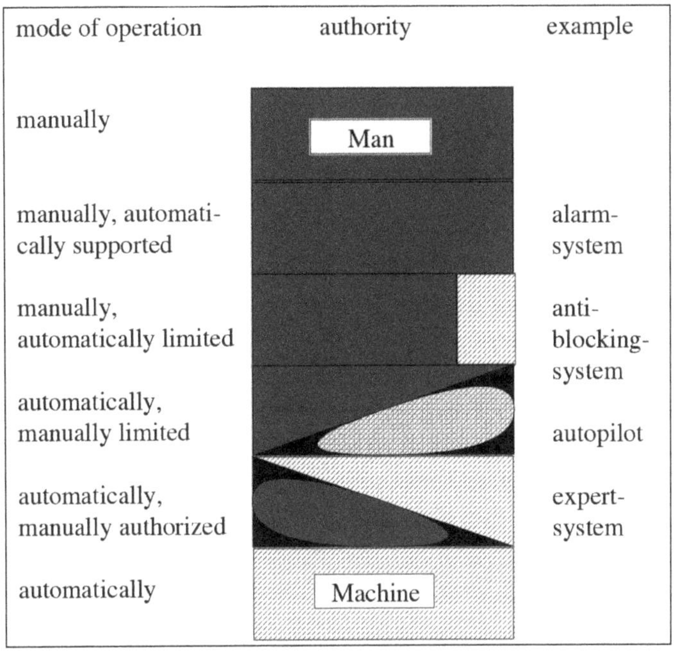

Different levels for the allocation of work authority between man and machine.

A Change of Paradigm in Robotics Objectives

Another concern of actual and future robot application is its socio-economic implications. Problems such as job availability for humans and the distribution of income in an "automated" society will have to be discussed. It appears to be most essential to design the relation between robot and human as co-operative and not as competitive. From the part of robotics the change of paradigm, instead of building machines that can do the work of humans, we should build machines that can do the work which humans cannot do, or do not want to do.

This would extend the overall range of human work, and it would mean that in most applications the robot can be considered to be an intelligent tool. For such a robot with a co-operative behaviour towards the human user the nice term "cobot" has been used recently. It is obvious that the classical definition of a robot will not cover these novel features of man-machine-interaction, and therefore a modification is introduced subsequently.

A New Definition of a Robot

One of the classical definitions for an industrial robot, as characterized in figure, says:

> "A robot is a freely and re-programmable, multifunctional manipulator with at least three independent axes, for moving materials, parts, tools, or special devices on programmed, variable trajectories in order to fulfill a multitude of various tasks."

This definition was practically adopted by the international ISO-Standard. But, in view of the said above, it is understandable that this definition does not meet the expectations and needs of future development. Formerly, there was no discussion about autonomy, mobility or even intelligence of robots. Therefore, other definitions have been suggested which take the relation to humans into account. A recent definition says:

> "Robots are sensumotoric machines for the extension of human mobility. They consist of mechatronic components, sensors and computer-based control functions. The complexity of a robot differs clearly from other machines by the larger number of degrees-of-freedom and the variety and the scope of its behaviours."

This modern class of machines is expected to behave in a more or less intelligent way. The term intelligence is deeply related to human beings. If some elements of this intelligence are transferred to other areas - biological, social, or technical ones, e.g. machine intelligence - it needs special definitions to show the limits and differences. The term robot intelligence may be justified by its usefulness to characterise some very desirable, human-like features, to be implemented in robots.

Legal and Ethical Aspects and Challenges

Beyond the technical, ergonomic and socio-economic aspects, basic questions as to the role of machines in human life, leading to legal and ethical issues, have come up.

Legal Challenges

When robots become more and more part of daily life, legal questions of their interactions with humans and their consequences have to be addressed. At the time being, robots are considered machines, and liability is being judged by existing regulations. In Europe, for example, machines are required to have a quality standard that corresponds to the state of the art and is defined by the machine guidelines, and in addition there is the well-known product liability.

The risk potential for robots that move around freely, in the same work space as humans, is much more difficult to classify. Intelligent robots are supposed to be able to learn, and to adapt their behaviour to unstructured and changing environments. As an example we can look at a robot transport assistant, helping an airport passenger to bring his luggage from the conveyor belt to the car or train, or a messenger robot delivering an express parcel within an office building or even across the city. It is easily conceivable that the exceptional situations during such a robot task cannot be completely predicted, that malfunctions cannot be excluded, and that the risk of humans being injured by such a robot is immanent. Who would be responsible? Potential addressees are: the owner of the robot, the operator, the producer, the person who has written the learning programm, the teacher who has taught the robot, the injured person himself, or a third party. Certainly, the use of such an "intelligent" robot would require some kind of insurance coverage, and in addition, there will have to be technical means to clarify any critical situation: the robot has to be identifiable, and it will have a memory that is accessible to document the past life of the robot. A detailed study of the legal questions coming up with the use of intelligent robots in various application areas, in particular for medicine, is given by Schweighofer.

Philosophical and Ethical Issues

Sturma states that robotics has more consequences for our personal and cultural self-understanding than other technologies. In robots, man appears to meet himself. Variations of such encounters can be found in myths and fictions about the artificial man, which go back to antique Greek times.

Sturma pledges, among others, that the unjustified use of anthropological terms such as 'conscience', 'knowledge', 'way of life', 'intention', 'autonomy', 'behaviour' can lead to a failure of appreciation of basic differences between man and machine and to problematic lines of reasoning. For example, it appears to be rather popular to claim in a cultural pessimism that in future humans might be replaced by robots or computers, be it for the better or the worse.

The ideas about what machines could or should do to make them "behave" in a complementary or even similar way to human beings touch upon a broad range of human values. As an example, there are ethical rules such as the famous "Three Laws of Robotics" from Asimov's utopian novels. As a reminder, Asimov's three laws run as follows:

- A robot may not injure a human being, or, through inaction, allow a human being to come to harm.
- A robot must obey the orders given it by human beings except where such orders would conflict with the First Law.
- A robot must protect its own existence as long as such protection does not conflict with the First or Second Law.

Of course, such rules cannot be implemented technically, even in near future. However, they can be regarded as a contribution to develop orientational knowledge in addition to the technology dominated factual knowledge that is necessary for actually building robots. The time is ripe for a step towards considerations of this kind, and who else would be able to recognise the implications and to acknowledge at least partially the responsibility if not the involved scientists themselves? Ethical rules would not mean restraints on basic science - once a question has been asked it cannot be made unthought and eventually someone will seek an answer to it – but they certainly would put certain constraints on applications. It is quite obvious that intelligent robots can be applied in very sensitive areas, such as medical care, military, or biology and not everything that can be done should be done. At least, any responsible engineer should develop his or her individual value judgement and be able to make accountable decisions.

Applications and Examples

Intelligent robots will offer novel chances in various ways and for different areas. The subsequent selection of applications gives some comments on the chances offered:

- Robots for the extension of the human work range are being used in areas such as space, underwater, and the micro- and nano-world.
- Robots for the alleviation of humans from hard or dangerous work will be used as intelligent tools in complex environments. Examples are machinery for construction work, in tunnelling, sewage channelling and cleaning, waste disposal, de-mining, de-construction of nuclear power-plants.
- Service robots, serving humans for making life easier, will be built for delivery services in office environments and hospitals, cleaning, or lawn mowing. An economic profit is difficult to assess, as they are most often directly competing with relatively cheap and still somewhat attractive human labor.
- Edutainment-robots (for education and entertainment) will probably be the first ones to introduce advanced concepts of "intelligence". They have the advantage that malfunctions usually do not endanger the user and are of no serious economic consequences, and that they are therefore very suitable to try out novel ideas, even in mass production. Playing football with robots is just one of the most popular topics in this field. It is typical that computer games are increasingly integrating artificial intelligence into their programmes as demonstrated for example by the game "Republic – The Revolution".
- Medical techniques are a dominant research area in robotics today. It includes prostheses for limbs and hands, artificial organs for audio and vision sensing, other techno-implants, surgery robot assistants, in particular for endoscopic surgery, robot helping assistants for handicapped, for rehabilitation, or for the aged to enable them to stay in their familiar environment as long as possible.
- Embedded robotics is an extension of machine-oriented robotics to systems. Applications range from 'cartronics', 'domotronics', to medical systems, and to advanced man-machine-interfaces making use of wearable computing equipment.

Nano-Robot as an Example for the Extension of Human Work Range

Handling of very small particles with nano-precision has been the domain of physicists. Exploiting and using the results on an industrial scale needs automated processes, it needs the transfer of technology known from robotics. Figure shows the set-up of a nano-robot-system. A stereo vision module and other sensors are used to localise the gripper and the object. This information is transmitted to the vision-based controller for generating automated robot motions. Additionally, the same information is used to create a virtual reality scene to allow the operator to observe the situation and to manually interact.

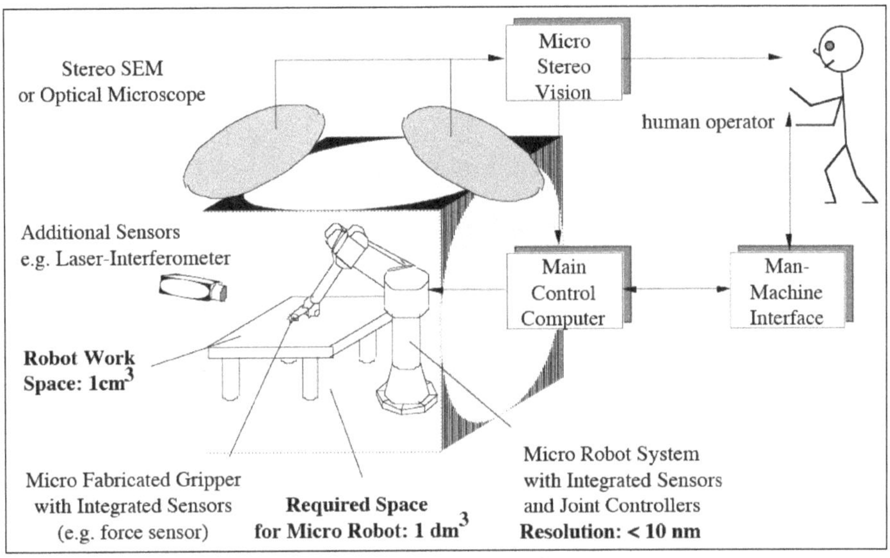

Overview on the ETH nano-robot-system.

Intelligent Control of Leg Prosthesis as an Example for Medical Applications

The above-knee prosthesis shows the direct interaction between man and machine. The prosthesis can be seen as a robot that is, at least partially, replacing human walking functionality. The control of the rather complex motion pattern is based on a finite set of rules. The pre-defined and stored parameters, models, and rules are combined, forming the knowledge base of the system. The soft control is algorithmic and composed of four sub-functions. The first task is gait state identification based on online measurements. Then the gait control computes the damping levels of the knee motion depending on the various gait phases. In order to realise that impedance control, an actuator opens and closes a valve in the hydraulic actuator connected to the knee joint. Thus the energy for controlling the impedance is very small; the major part of the energy for walking is still coming from the motion of the hip. Finally, the gait speed is identified, a parameter which influences both the gait control and the hydraulic control. A laboratory prototype of such a controlled prosthesis was built and evaluated in a rehabilitation clinic, showing a number of distinct advantages over classical prostheses. Figure points to a success story. The potential of the novel prosthesis is mainly based on its ability for internal information processing, and, in future, it can still be enhanced by making use of learning algorithms to adapt the prosthesis to the individual gaits of its bearer.

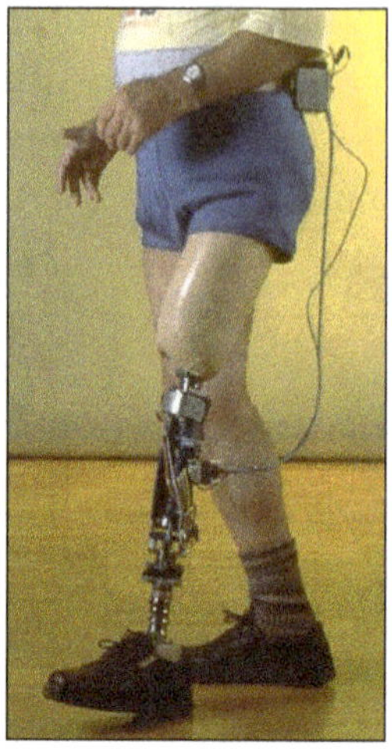

Intelligently controlled above-knee prosthesis in clinical test.
The prototype computer is fastened to the belt.

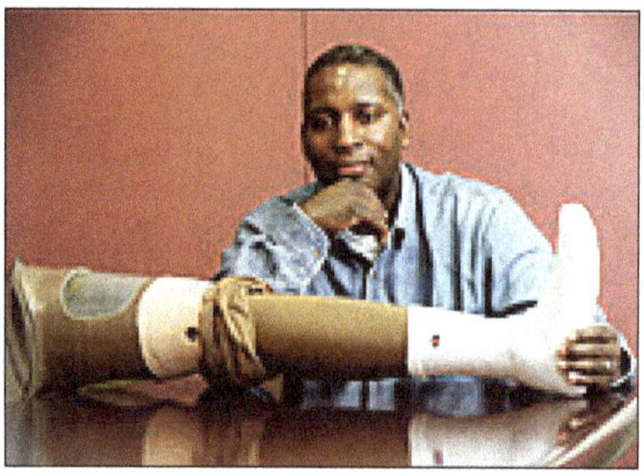

LIFESAVER - Curtis Grimsley used his Otto Bock C-Leg to walk down from the 70th floor of the World Trade Center on Sept. 11. He says the device saved his life.

Medical Robot Systems

Obviously, in medical robot systems the interaction between man and machine is very close, giving rise to a number of questions, ranging from technical to ethical ones. Medical robots for surgery on the hip even open heart, and supporting tools for minimal invasive surgery are well known, already, and are under continuous development. Figure shows a simulated setup for robot assisted surgery. The KISMET (Kinematic Simulation, Monitoring and Off-Line Programming Environment for Telerobotics) software is under development at Forschungszentrum Karlsruhe since 1986 as a 3D realtime simulation support tool for medical applications. Research on such medical robot systems is being done at the Robotics- and Mechatronics Lab of the DLR.

One of the most powerful tools for advancing medical techniques actually is the integration of computers. For example, the fundamental target of a Swiss national research program is to understand, realise, and demonstrate the potential, which information technology offers for the optimisation of medical interventions in order to improve both the treatment of individual patients and the overall health care for the whole society. The focus of this program is on the development, integration and validation of enabling technologies towards advanced computer aided and image guided systems for medical interventions that support the complete treatment process from therapeutic planning and simulation via intra-operative action to postoperative care, monitoring and documentation.

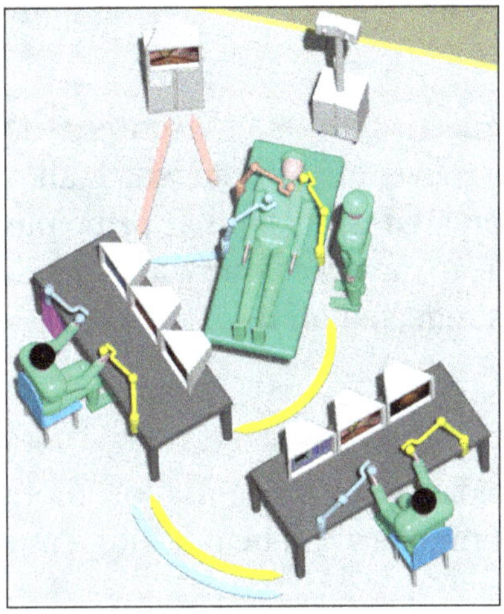

Robot assisted minimal invasive surgery of the future.

Biomimetic Robotics

Humans have been inspired by nature's naturally evolved creations and have been mimicking it for many years. Both Leonardo DaVinci and the Wright brothers took their flying inspiration from birds and the creator of Velcro, George de Mestral, was inspired by the hooks of burseeds. Since then advances in both robotics and artificial intelligence have allowed the field of biomimetic robotics to develop. Applying biomimicry principles to our technologies allows us to understand, adapt and solve complex control, manipulation and manoeuvrability problems. While biomimetic robotics spans over a large number of fields.

Types of Mimicry

Within the field of biomimicry, there are a numerous of areas of interest. A number of these areas are discussed below:

Material

Biomimetic materials can be categorised into four different groups:

- Smart materials inspired by nature's ability to react and change in response to external stimuli.

- Surface modifications which include novel surface topographies with improved functions.
- Material architectures which feature novel shapes and structural arrangements.
- Technologies which are based on enhancing existing systems using specific parameters of an adaptation (rather than developing an entirely new technological platform).

One example of a material reacting to its environment is evident in butterfly wings. Butterfly wings allow light to reflect off of them which allow them to change colour. An iridescent coating has been created using similar principles which allow vehicles, such as some Range Rover models, to have a colour which does not fade and changes colour depending on how the sun is reflected off it.

Locomotion

Biomimicry also has a major impact on robotic locomotion and kinematics. As with materials, the locomotive biomimicry can be put into a number of categories:

- Improvement based on movement kinetics.
- Improvement based release mechanisms (means of dispersal across an environment).
- Improvements based on structural configuration (energy efficient shapes).

Inspiration for mimicry can come from both animals and humans.

Animals

A lot of research has gone into trying to figure out how different animals travel. These include animals from all three physical environments: land, sea and air. Animals such as fish and squids, to two-, four- and six- legs robots have been investigated.

Soft Robotics

One form of biomimetic robotics that has recently advanced by mimicking animal materials is the field of soft robotics. Soft robotics allows for soft and deformable robotics. This material is already being used in a number of robots.

Festo AquaJellies 2.0.

The propulsion of this soft robot is controlled by alternating between tension and pressure flanks. This causes the robots 'tentacles' to move in a wave-like motion, bending when under tension and then relaxing. This is very similar to the way jellyfish move in real life. Additionally, researchers at Harvard University have created a robot that is inspired by animals that lack internal skeletons such as squids and starfish.

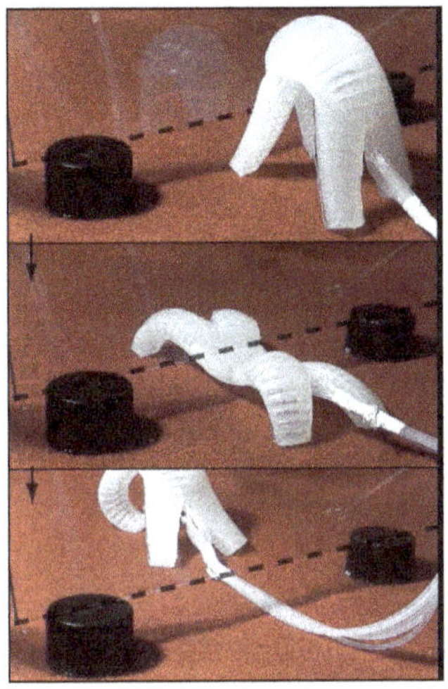

Harvard University tetrapod.

This robot uses a mixture of crawling and undulating gaits. As with the 'AquaJelly' the robot flexes when pressurised. As the robots structure is separated into a number of compartments, different areas can be pressurised allowing different movements. The soft material of the robot allows it to move underneath an obstacle as can be seen in figure. This task would be almost impossible for a hard robot. Soft robots tend to be very flexible and this makes them very resilient and hard to damage.

Birds

The biomimicry of birds has been around ever since humans wanted to fly. Travelling through air and water share similar properties. In the air however thrust has to be generated. Festo created both marine and air biomimetic robotics and filled the air robots with helium.

Festo AquaPenguin.

Festo AirPenguin.

Another robot that has been designed to mimic a bird is the RobotSwift. The RoboSwift is a micro aeroplane with moveable wings. It has a streamlined design with allows it's to adjust the wing area and camber in the same way a swift would.

Insects

Insects also have many properties that humans have learnt a lot from. The locomotion of cockroaches has been investigated during the development of "iSprawl". The iSprawl, like a cockroach, has 6 legs and by modelling it on the cockroach allows it to investigate natural locomotion dynamics and the optimal arrangement of 6 legs. Another robot inspired by insects is Boson Dynamics' 'Sandflea'.

Boston Dynamics Sandflea.

The 'Sandflea' robot mimics the way in which fleas can jump and this robot can jump an impressive 1-8 metres. It is extremely robust and water resistant, allowing it to recover well from falls undamaged.

Mammals

The gait of much larger mammals has also been researched. Boston Dynamics' 'Cheetah' robot is based off the cheetah which is the fastest land animal. By mimicking this animal, it has allowed the creation of a robot that can travel 29 mph which is a land speed record for legged robots.

Another biomimetic mammal that has been mimicked is the Kangaroo. Kangaroos are well known for their ability to jump and stabilise themselves. Festo's BionicKanagaroo is able to store the energy of its jumps just like a real kangaroo.

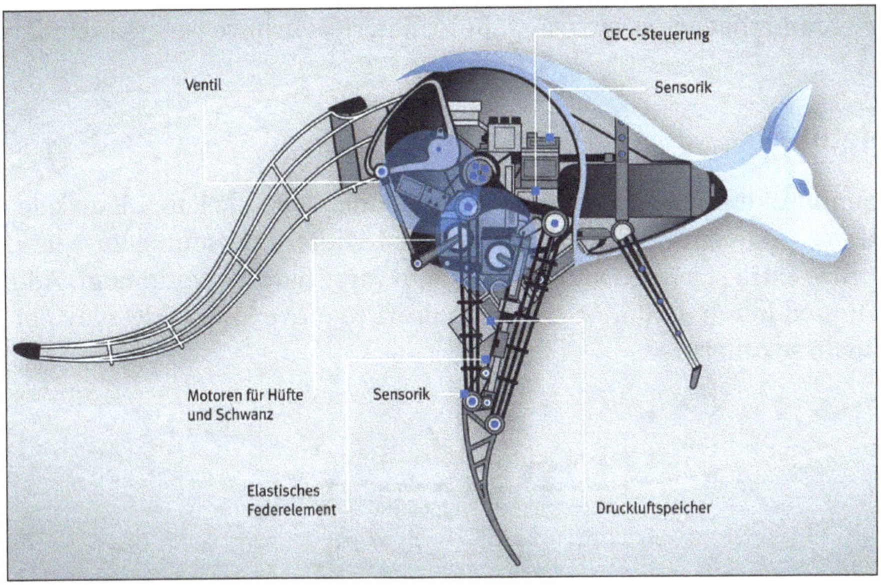

Festo BionicKangaroo.

Human Limbs

As well as mimicking animals to learn and develop our technologies it has also be beneficial learning from our own bodies. There are many people who have physical disabilities who may be aided by biomimetic technologies.

Bionic Limb

Human hands are often mimicked in robotics due to curiosity over its complexity. There are a number of advantages to having a bioinspired hand rather than a gripper. For example, by increasing the number of degrees of freedom it allows the robot to grasp a wider variety of shaped objects. Additionally, by mimicking the human hand it allows the user to fully control it in a more intuitive manner.

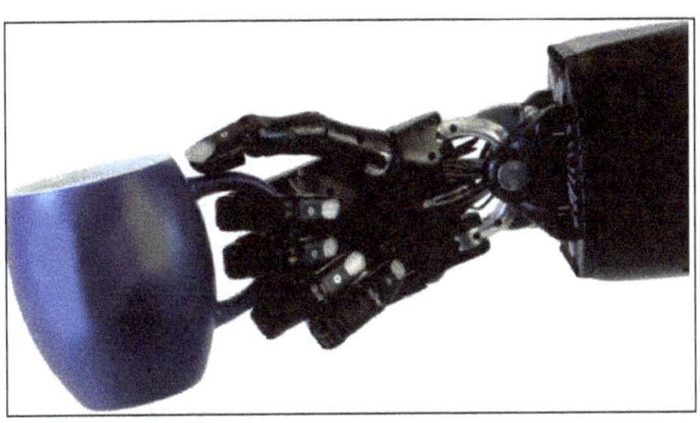

ShadowRobotics Dexterous Hand.

Artificial Muscles

Another thing that has aided the creation of biomimetic limbs is the development of artificial muscles. The 'muscles' are mimicked using Electro-active Polymers actuators. It has been found that actuators using these materials behave very similar to biological muscles.

Exoskeletons

A final example of human based biomimetic robotics is exoskeleton. Exoskeletons have been used in a number of different applications. Military personnel have used them in order to prove extra strength, allowing them to carry heavier equipment. Additionally, it has been used in rehabilitation, allowing users to move joints they may not have the strength to do so otherwise.

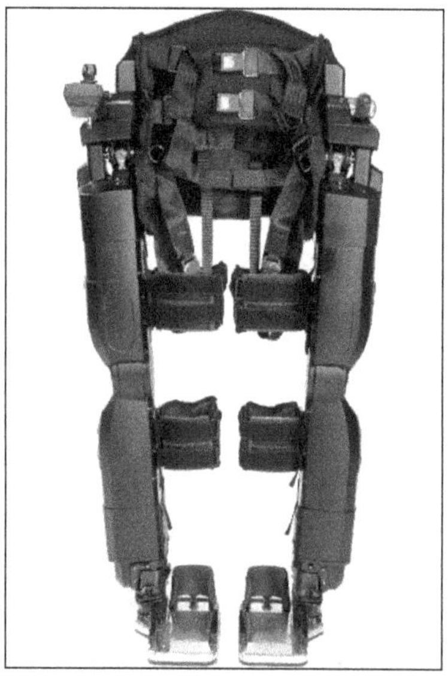

RexBionic.

Behaviors

As well as looking at the physical properties of nature there is also a lot to learn from their behavior. A single ant on its own is limited by what it achieves, however when a number of ants are put together they act as a unit which is known as a swarm. By implementing this swarming behavior within robots it allows robots to work together to achieve a much larger and more complicated task which would be impossible to achieve alone.

Applications

Biomimetic robotics is used in a number of different applications. These range from medicine to space to the defence sector. In all these fields, nature has provided a way of improving the required technologies.

Medicine and Surgery

Biomimetic robotics plays a large role in the creation of bionic limbs and exoskeletons. Assistive biologically inspired robots provide an important rehabilitation role. Another way in which nature can help aid medicine is in relation to drug delivery technology. Animals such as sea anemones and corals have stinging cells that can inject the skin. This is particularly beneficial for transdermal drug delivery as it would allow for the skin barrier to be penetrated.

Soft robotics can also be used to assist minimally invasive surgeries. Using an octopus inspired robot allows the soft instrument to reach areas of the body that solid instruments may struggle to reach. Additionally, due to soft robots having the ability to stiffen varying parts, there is the possibility of it being able to hold organs out the way while working on another part of the body.

Automobile

Biomimicry is also used in the automotive industry. With the increased research in self-driving cars, the need for reliable collision avoidance systems has increased. In 2008, Nissan engineers looked at the behaviour of bumblebees in order to tackle this problem as bumblebees are well known for avoiding obstacles when travelling at speed.

Military

The military have also invested in a lot of money towards biomimetic robotics. Boston Dynamics' 'Big Dog' robot is a rough terrain robot using animal-like mobility.

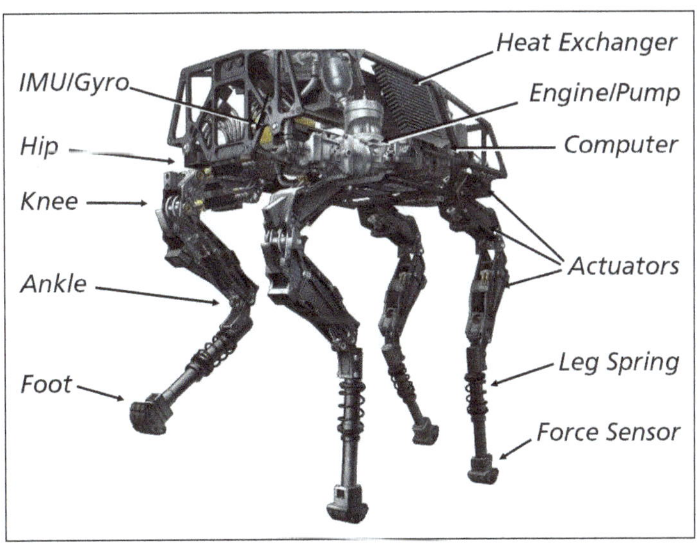

Boston Dynamic Big Dog.

'Big Dog' can access areas where wheeled and tracked robot cannot go allowing it to acts a way of carrying bags for the military.

Space

It is noted that biomimetic robotics can also prove beneficial in space explorations. By imitating the chimpanzees knuckle walking it would allow for the creation of a robot that could walk on two and four feet thereby increasing the effectiveness when exploring difficult terrain.

Swarm Robotics

Swarm robotics is a novel approach to the coordination of large numbers of relatively simple robots which takes its inspiration from social insects. This paper proposes a definition to this newly emerging approach by:

- Describing the desirable properties of swarm robotic systems, as observed in the system-level functioning of social insects.
- Proposing a definition for the term swarm robotics, and putting forward a set of criteria that can be used to distinguish swarm robotics research from other multi-robot studies.
- Providing a review of some studies which can act as sources of inspiration, and a list of promising domains for the utilization of swarm robotic systems.

Swarm robotics is a novel approach to the coordination of large numbers of robots. It is inspired from the observation of social insects —ants, termites, wasps and bees— which stand as fascinating examples of how a large number of simple individuals can interact to create collectively intelligent systems. Social insects are known to coordinate their actions to accomplish tasks that are beyond the capabilities of a single individual: termites build large and complex mounds, army ants organize impressive foraging raids, and ants can collectively carry large preys. Such coordination capabilities are still beyond the reach of current multi-robot systems.

Motivations for Swarm Robotics

Studies have revealed that there exists no centralized coordination mechanisms behind the synchronized operation of social insects, yet their system-level functioning is robust, flexible and scalable. Such properties are acknowledged to be desirable for also multi-robot systems, and can be stated as motivations for the swarm robotics approach:

- Robustness requires that the swarm robotic system should be able to continue to operate, although at a lower performance, despite failures in the individuals, or disturbances in the environment. As anyone who tried to extinguish an ant raid into his kitchen would agree, social insects are extremely difficult to get rid of. This robustness can be attributed to several factors; First, redundancy in the system; that is, any loss or malfunction of an individual can be compensated by another one. This makes the individuals dispensible. Second, decentralized coordination; that is, destroying a certain part of the system will not deter the system's operation. Coordination is an emergent property of the whole system. Third, simplicity of the individuals; that is, in comparison to a single complex system that could perform the same task, in a swarm robotic system, individuals would be simpler, making them less prone to failures. Fourth, multiplicity of sensing; that is, distributed sensing by large numbers of individuals can increase the total signal-to-noise ratio of the system.

- Flexibility requires the swarm robotic system to have the ability to generate modularized solutions to different tasks. As nicely demonstrated by ants, in ant colonies individuals take part in tasks of very different nature such as foraging, prey retrieval and chain formation. During the foraging task, ants act independently searching for food in the environment; their search is partially coordinated by the pheromones laid in the environment. The prey retrieval task requires the ants to generate a force much larger than that of a single individual to drag a prey to the nest. When a large prey is discovered, each ant grips the prey with its mandible and pulls it in different directions. The seemingly random pulls of ants are observed to be coordinated through the force integrated over the prey. In the chain formation task, ants form a physical chain-like structure that can extend beyond the reach of a single ant and exert large forces pulling together leaves. During the task, ants use their body as a medium of communication where ants in the chain act motionless with each ant gripping/holding the leg of other ants in the chain. In this task, coordination is achieved through the bodies of the ants. Swarm robotic systems should also have the flexibility to offer solutions to the tasks at hand by utilizing different coordination strategies in response to the changes in the environment.
- Scalability requires that a swarm robotic system should be able to operate under a wide range of group sizes. That is, the coordination mechanisms that ensure the operation of the swarm should be relatively undisturbed by changes in the group sizes.

Swarm Robotics

The term swarm intelligence was first coined by Gerardo Beni as a "buzz word" to denote a class of cellular robotic systems. However, the term was embraced more by the social insect studies and by the optimization studies that used the social insect metaphor, losing much of its original robotics context. During recent years, the term swarm robotics emerged as the application of swarm intelligence to multi-robot systems, with emphases on physical embodiment of the entities and realistic interactions among the entities and between the entities and the environment. In a sense, the term swarm robotics took the heir of swarm intelligence which moved on to cover a broader meaning.

Although, like every other newly coined term, swarm robotics will have a life of its own to claim its meaning, our observations indicate that such new terms run the risk of turning into buzz words that tend to be attached to existing approaches with little thought over whether it really fits or not. Such misuses, in time, can drift the term in every direction blurring the very point that made it novel. In an attempt to prevent this, we will propose a definition and a set of distinguishing criteria for the swarm robotics approach.

As our starting point, we propose the following definition for the term swarm robotics: Swarm robotics is the study of how large number of relatively simple physically embodied agents can be designed such that a desired collective behavior emerges from the local interactions among agents and between the agents and the environment.

This definition by itself, however, is not sufficient to properly describe this newly emerging term. Within the multi-robot research only, there already is a plethora of terms labeling different flavors of multi-robot research such as "collective robotics", "distributed robotics", "robot colonies", with often vague and overlapping meanings. Therefore, we would like to put forward a set of criteria for distinguishing swarm robotics research.

Autonomous Robots

As much as it seems obvious, we believe that the requirement that the individuals that make up the swarm robotic system be autonomous robots needs to be explicitly stated. That is, the individuals should have a physical embodiment in the world, be situated, can physically interact with the world and be autonomous. Sensor networks that consist of distributed sensing elements, but with no physical actuation abilities, should not be considered as swarm robotic systems. Yet we believe that the studies on sensor networks are highly relevant for swarm robotics.

The metamorphic robotic systems, in which units adhere to each other and can, only, move over each other by forming and disconnecting connections with other units can also be considered as swarm robotic systems as long as there exist no centralized planning and control centers.

Large Number of Robots

The study should be relevant for the coordination of a "swarm of robots". Therefore, studies that are applicable to the control of only a small number of robots and do not aim for scalability, fall outside swarm robotics. Although putting a number as a lower bound of group size is difficult to justify, and most would accept group sizes of 10-20 as "swarms". Despite the lowering cost of robots, maintenance and experimentation with large groups of robots will remain as a main obstacle. Therefore the issue of relevancy is mentioned to express that the field should be open to studies that are carried out with smaller group sizes, but with the vision/promise of scalability in sight.

Few Homogenous Groups of Robots

The robotic system being studied should consist of relatively few homogeneous groups of robots, and the number of robots in each group should be large. That is, studies that are concerned with highly heterogeneous robot groups, no matter how large the group is, are considered to be less "swarm robotic". For instance, studies on robosoccer teams mostly fall outside of swarm robotics since these teams typically consist of individuals whose different "roles" are assigned to them by an external agent prior to the operation of the team and hence they are highly heterogeneous.

We agree that, the issue of homogeneity in a group of robots is not a trivial one. In Balch proposed a metric, called the hierarchical social entropy, which can be used for this purpose. Yet, it is difficult to determine whether two individuals belong to the same group or not using a simple evaluation run in the evaluation chamber. This is due to two reasons: 1) the nonlinear inter-robot interactions will have a large effect on the behavior of the robots, and 2) probabilistic behaviors can make it impossible to obtain exact similar evaluation runs under exactly the same conditions.

Relatively Incapable or Inefficient Robots

The robots being used in the study should be relatively incapable or inefficient on their own with respect to the task at hand. That is, either: 1) the robots should have difficulties in carying out the task on their own, and the cooperation of a group of robots should be essential, or 2) the deployment of a group of robots should improve the performance/robustness of the handling of the task. Collective retrieval of a large prey by ants is a good example to the first case where retrieval by a single ant would be impossible. Collective foraging of ants using pheromones laid on the ground for stigmergic communication creates foraging patterns which are believed to improve their foraging performance. Using a group of simple mobile robots, Sugawara et al showed that signalling the discovery of an object in environments where objects are non-uniformly distributed can yield super-linear increases in the performance of the swarm.

It is important to note that this criterion does not impose any restrictions on the hardware and software complexity of the robots. The incapability and inefficiency of individual robots should not be taken in absolute terms; rather they should be seen relative to the task and be considered as a justification for the simplicity of robots.

Robots with Local Sensing and Communication Capabilities

The robots being used in the study should only have local and limited sensing and communication abilities. This constraint ensures that the coordination between the robots is distributed. In fact, the use of global communication channels within the robot group is likely to result in unscalable coordination mechanisms and would therefore act against the first criterion. However, note that the global communication channels, which can be used as a means to download a common program onto the swarm, are acceptable, as long as it is not used for coordination among the robots.

Aerial Robotics

Beholding the industry trends, we can say that a substantial interest in aerial robotics has grown in recent years. Flying surely opens up new opportunities to robotics services to serve several tasks including search and rescue, observation, mapping and navigation along with inspection and maintenance. The technology of Aerial Robotics enhances the human-robotics collaborations for better.

The pace at which robotics industry is expanding is giving rise to new disruptive technologies which are transforming the existing operations in the industry. When the aerial technologies are blended with the virtue of robotics, Aerial Robotics enters the picture. The technology, in basic, is built to simplify the drone operations automating the digitization of sites and locations through high-frequency data collection, processing, visualization, and analysis.

Without a doubt, the basic structure of aerial robots, unmanned aerial vehicle, have foundations built in aircraft and automatic flight control system. Notably, the first UAV ever made was based on the modified structure of manned rockets and aircraft architecture. The evolution of such technology has always challenged the disruption leading to miniaturization, the agility of flight control, perceptual skills, navigational autonomy and flight endurance along with aerial manipulation.

Current Trends

The concept of system miniaturization, agile flight control, perception, and 3D mapping, autonomous motion planning, Multi-Robot Systems, Augmented Human-Robot and Prolonged Endurance, merged with the unmanned aerial vehicle, and is disrupting the prevailing aircraft market for sure. The current trends of aerial robotics booming the industry are:

- Retaining stability during propeller failure.
- Convertible unmanned aerial vehicle.
- Aerial mapping.
- Information Gain-based efficient exploration and 3D mapping of unknown environments.
- Localizability-aware autonomous exploration and mapping.
- Multi-Modal localization and mapping.
- GPS-denied navigation.

The research and development of unmanned aerial vehicle have accelerated at a fast pace in the past decade.

Impact on Society

With new technology comes a heap of responsibilities not only towards business and development but also considering its social impact. Every technology that ever existed has influenced society in certain ways mostly proving itself as a boon to mankind. The similar convention goes with aerial robotics too. The technology surely has some indicative applications for the betterment and welfare of people.

Aerial robotics contributes to infrastructure inspection and maintenance of different architectures. The technology also provides with humanitarian assistance. Precision agriculture, based on observing, measuring and responding to inter and intra-field variability in crops as a part of farming management, is also a major beneficiary of this technology. Along with that, aerial robotics is also helpful in the climate control process.

Above all, the most significant deployment of aerial robotics in the welfare of humankind is security and surveillance. In fact, police and defense forces use drone and other aerial equipment to ensure the security and safety of citizen.

Challenges

The technology of aerial robotics is definitely a great approach to change the world and bring about significant transformation in society. Regardless of what it has achieved today, every technology and its application has to go through different challenging phases which tests their boundaries and odes that not even the sky is the limit for them.

Aerial robotics technology also possesses some vital challenges which need to overcome with time and innovation. This robotics technology is reliable for sensible decision making in the world of intelligent automation or not. Certain other concerns are also creating a buzz around the corner.

- If any particular skill set is required for operating aerial robots?
- Is there any future for aerial robotics technology beyond flying camera?
- If the technology worth assigning complex tasks for execution or not?
- What are the collision avoidance features of aerial robotics?
- In controlled national airspace, will it be trustworthy to fly aerial robot in urban areas?

The industry is already dealing with the talent gap and security breaches in various tech-adoptions. Amidst this, the application of aerial robotics adds to the concerns of experts and professionals regarding its potential uses and misuses. This simply implies that future aerial robots have to deal intelligently with uncertainties in their path.

Cloud Robotics

Robotic systems have brought significant socioeconomic impacts to human lives over the past few decades. For example, industrial robots (especially robot manipulators) have been widely deployed in factories to do tedious, repetitive, or dangerous tasks, such as assembly, painting, packaging, and welding. These preprogrammed robots have been very successful in industrial applications due to their high endurance, speed, and precision in structured factory environments. To extend the functional range of these robots or to deploy them in unstructured environments, robotic technologies are integrated with network technologies to foster the emergence of networked robotics.

A networked robotic system refers to a group of robotic devices that are connected via a wired and/or wireless communication network. Networked robotics applications can be classified as either teleoperated robots or multi-robot systems. In the former case, a human operator controls or manipulates a robot at a distance by sending commands and receiving measurements via the communication network. Application examples include remote control of a planetary rover and remote medical surgery.

In the latter case, a team of networked robots complete a task cooperatively in a distributed fashion by exchanging sensing data and information via the communication network. Examples include cooperative robot manipulators, a team of networked robots performing search and rescue missions, and a group of micro satellites working cooperatively in a desired formation.

Networked robotics, similar to standalone robots, faces inherent physical constraints as all computations are conducted onboard the robots, which have limited computing capabilities. Information access is also restricted to the collective storage of the network. With the rapid advancement of wireless communications and recent innovations in cloud computing technologies, some of these constraints can be overcome through the concept of cloud robotics, leading to more intelligent, efficient and yet cheaper robotic networks. In this topic, we describe cloud robotics architecture, some of the technical challenges, and its potential applications. Some preliminary results on the optimal operation of cloud robotics are also presented.

Challenges in Networked Robotics

Networked robotics, especially the multi-robot system as illustrated in figure, distributes the workload of sensing, actuating, communication, and computation among a group of participating robots. It has achieved great success in industrial applications, intelligent transportation systems, and security applications. However, the advancement of networked robotics is restricted by resource, information, and communication constraints inherent in the existing framework.

Resource Constraints

Although a robot can share its computation workload with other units in the formation, the overall effectiveness of the robotic network is limited by the collection of each robot's resources, including onboard computers or embedded computing units, memories, and storage space. Physically, these onboard computing devices are restricted by the robots' size, shape, power supply, motion mode, and working environment. Once the robots are designed, built and deployed, it is technically challenging, if not infeasible, to change or upgrade their resource configurations.

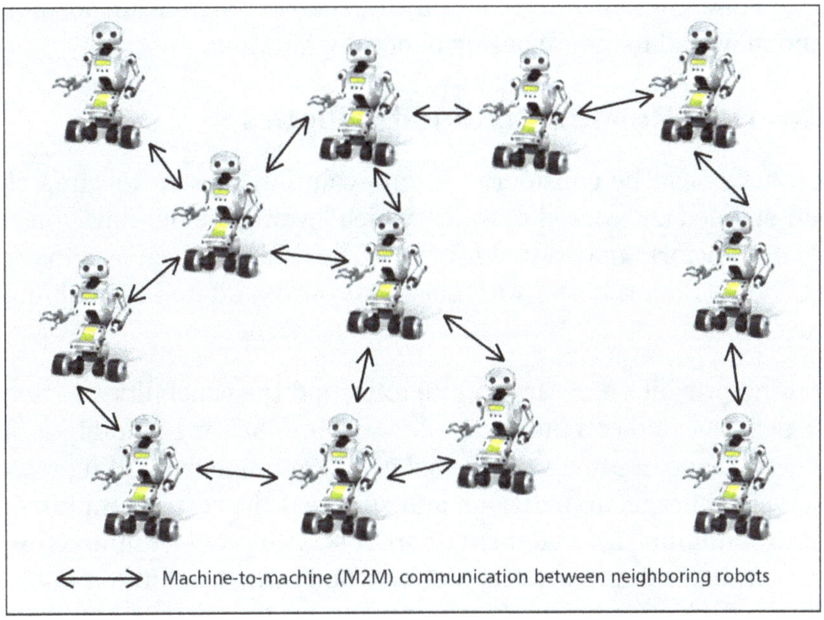

Networked Robotics: a team of robots are interconnected with a communication network to collaboratively accomplish tasks.

Information and Learning Constraints

The amount of information a robot has access to be constrained by its processing power, storage space, and the number and type of sensors it carries. Networked robotics allows the sharing of information amongst robots connected by a communication network so that a global task can be solved or computed cooperatively using the whole network. However, networked robotics is constrained by the information observed or computed by robots in the network, and by the examples or scenarios that the network encounters, and hence limiting its ability to learn. A robotic team learning to navigate may perform very well in a static environment, where all obstacles can be mapped out with increasing accuracy over time. On the other hand, the learning process has to be repeated once the environment changes or the robotic team is placed in a new unfamiliar environment. The map databases maintained by the robotic formation are also limited by the collective amount of storage space (including memory and disk) the formation has.

Communication Constraints

Common protocols for machine-to-machine (M2M) communications include proactive routing, which involves the periodic exchange of messages so that routes to every possible destination in the network are maintained, and ad-hoc routing, which forms a dynamic route to a destination node only when there is a message to be sent. Proactive routing incurs high computation and memory resources in the route discovery and maintenance process. Ad-hoc routing protocols suffer from a high latency as a route has to be established before a message can be sent, and are not practical if the network topology is highly dynamic. These drawbacks are significant in mobile robotic networks, and may lead to severe performance degradation.

From Networked Robotics to Cloud Robotics

Networked robotics can be considered as an evolutionary step towards cloud robotics, i.e., cloud-enabled networked robotics, which leverages emerging cloud computing technologies to transform networked robotics. The design objective is to overcome the limitations of networked robotics with elastic resources offered by a ubiquitous cloud infrastructure.

Cloud computing provides a natural venue to extend the capabilities of networked robotics. NIST defines cloud computing as "a model for enabling ubiquitous, convenient, on-demand network access to a shared pool of configurable computing resources (e.g., networks, servers, storage, applications and services) that can be rapidly provisioned and released with minimal management effort or service provider interaction." Through its three service models (i.e., software, platform and infrastructure), it enables tremendous flexibility in designing and implementing new applications for networked robotics.

Cloud Robotics

We first describe system architecture for cloud robotics, and then focus on the two key enabling subsystems: the M2M/M2C communication framework and the elastic computing architecture. Our cloud robotics differentiates from existing solutions in that it leverages two complementary clouds (i.e., an ad-hoc cloud and an infrastructure cloud).

System Architecture

In figure, we illustrate the system architecture for our proposed cloud robotics. The architecture is organized into two complementary tiers: a machine-to-machine (M2M) level and a machine-to-cloud (M2C) level.

On the M2M level, a group of robots communicate via wireless links to form a collaborative computing fabric (i.e., an ad-hoc cloud). The benefits of forming a collaborative computing fabric are multi-fold. First, the computing capability from individual robots can be pooled together to form a virtual ad-hoc cloud infrastructure. Second, among the collaborative computing units, information can be exchanged for collaborative decision making in various robot-related applications. Finally, it allows robots that are not within communication range of a cloud access point to access information stored in the cloud infrastructure or send computational requests to the cloud.

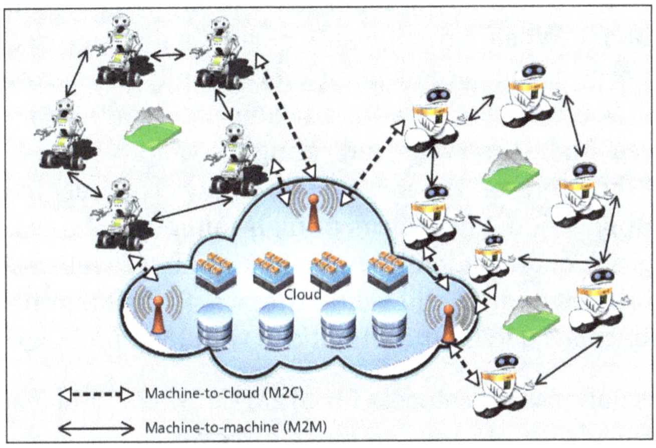

Cloud Robotics: robots are interconnected via M2M/M2C communications, sharing their resources and accessing to remote cloud resources.

On the M2C level, the infrastructure cloud provides a pool of shared computation and storage resources that can be allocated elastically for real-time demand. This elastic computing model allows the group of networked robots to offload computation-intensive tasks for remote execution, resulting in "remote-brain" robots. Moreover, the benefits of a large volume of storage provided by the centralized cloud are two-fold. First, it can unify a large volume of information about the environment, which can be organized in a format usable by robots. Second, it can provide an extensive library of skills or behaviors that are related to task requirements and situational complexities, making it feasible to learn from the history of all cloud-enabled robots.

M2M/M2C Communication Architecture

Robots in a network can communicate if they are within communication range of each other, and with the cloud servers if the robots are close to access points of the cloud infrastructure. A wireless M2M communication network can be formed by robots working cooperatively with each other to route and relay information. We call either a robot or an access point a node in the M2M/M2C network.

Several standards like Zigbee, Bluetooth, and WiFi Direct have been developed for short range wireless communications between robots. For long range communications, radio frequency and microwave communication technologies may be used.

A network of robots is often formed dynamically and in an ad-hoc manner. There is no central controller to coordinate the communication flow in the network. Robots may leave and join the network, or may become unavailable because of unpredictable failures or obstructions in the environment. Furthermore, the network is highly dynamic if robots are mobile. All these considerations make the design of effective routing protocols difficult and impractical in some scenarios. Gossip algorithms are randomized methods designed to transmit a message from a source to a destination without any explicit route discovery mechanisms. If two nodes are within communication range, we say that they are neighbors. When a robot wants to send a message to a destination node (either another robot or a cloud access point), it randomly chooses one of its neighbors and transmits the message together with a header that contains the identifier of the destination and itself, and a time value indicating the validity period of the message. In another variant of the protocol, the message is simply broadcast to all neighbors, but depending on the application, this may incur high communication load in the network. At every time step, each node randomly chooses a neighbor to retransmit messages that are not intended for itself and are still valid. After a sufficient number of time steps, all the messages will be relayed to their destinations with high probability.

We propose the use of gossip protocols for M2M/M2C communications in cloud robotics. Gossip protocols do not require route discoveries and maintenance, and are thus suited for highly dynamic mobile robotic networks. These protocols are also very simple to implement, and require minimal additional computation and memory resources. However, the trade-off is that gossiping may result in high message latency if the network conductance is low. In cloud robotics, however, this problem is significantly mitigated as the cloud backend infrastructure serves as a central super node for the M2M/M2C communication network. As we will see, the time required for a message to be disseminated in a network is greatly reduced by the existence of a super node. Alternatively, a hybrid gossip algorithm can be used in which routes to frequently accessed nodes like the group leader in the proxy-based cloud computing model, can be maintained.

Elastic Cloud Computing Architecture

Cloud robotics is built on the combination of an ad-hoc cloud formed by a group of networked robots and an infrastructure cloud. This unique combination offers us great flexibilities in designing computing models tailored for specific applications. Here we described the following three elastic computing models:

- Peer-Based Model: Each robot or virtual machine (VM) in the ubiquitous cloud is considered as a computing unit. These robots and VMs form a fully distributed computing mesh. A task can be divided into smaller modules for execution over a subset of the nodes in the computing mesh.
- Proxy-Based Model: In the group of networked robots, one unit functions as a group leader, communicating with a proxy VM in the cloud infrastructure, to bridge the interaction between the robotic network and the cloud. The set of computing units are organized into a two-tier hierarchy.

- Clone-Based Model: Each robot has a corresponding system level clone in the cloud. A task can be executed in the robot or in its clone. The set of robotic clones also form a peer-to-peer network with better connectivity than the physical ad-hoc M2M network. Moreover, this model allows for sporadic outage in the physical M2M network.

Elastic computing models for cloud robotics: a) peer-based model; b) proxy-based model; c) clone-based model.

Each of these elastic computing models exhibits different robustness in network connections, interoperability, and mobility flexibility. Robustness refers to the network connectivity between the set of networked robots and the infrastructure cloud. The clone-based model has the maximum number of linkages from robots to the infrastructure cloud, and is thus the most robust; the proxy-based model is the least robust in terms of network connectivity; and the peer-based model falls between these two extreme cases. Interoperability refers to the additional complexity required in operating a cloud robotics infrastructure with an existing robotic network. The proxy-based model is the most interoperable model, because of its hierarchical structure; while the clone-based model is the least interoperable. Mobility flexibility refers to the network's ability to support mobile robots. The peer-based model has the most flexibility for mobility, because VMs can be instantiated anywhere in the cloud infrastructure; while the clone-based model is the least flexible, because complicated VM migration mechanisms are required to support robot mobility.

Table: Comparisons of different computing models.

Model	Robustness	Interoperability	Mobility
Peer-based	Medium	Medium	High
Proxy-based	Low	High	Medium
Clone-based	High	Low	Low

The choice of a specific elastic computing model depends mainly on three factors, including network conditions, application requirements and resource availability. We aim to develop a unified framework to determine an optimal or near-optimal model for a given set of conditions.

Cognitive Robotics

Cognitive robotics is about doing robotics that deals with cognitive phenomena such as perception, attention, anticipation, planning, memory, learning, and reasoning. Now, some people believe that robotics already deals with those phenomena, and are therefore left wondering how cognitive robotics would be any different from robotics, period. However, despite what you see in the movies, most existing robots don't learn, have no memory to speak of, and don't reason. In fact, at this point most existing robots are used in industry (think assembly lines), and most of them don't even have any perceptual abilities at all; they are programmed to do one thing, and one thing only. This kind of robotics we might call Industrial Robotics, and it can be characterized with the 3 D's of robotics: robots that do dull, dangerous, or dirty work, that no human would or can do, which is exactly why Industrial Robotics is important! However, it is not what we see as Cognitive Robotics. In Cognitive Robotics, we are interested in the kind of robots that are, well, more cognitive. Robots with the kind of intelligence that humans have. Robots that reason, remember, learn, and that can communicate with humans and with each other. Robots that can be characterized by the 3 C's: Clever, Creative, and Charismatic.

2

Classification of Robots

The classification of robots can be done on the basis of various criteria such as the power source, workspace geometry, degrees of freedom, their application and environment. The topics elaborated in this chapter will help in gaining a better perspective about the classification of robots.

Classification based on Power Source

Electrical Power Source Robots

Robots operating with the electrical power source can further be subdivided as AC or DC systems. Direct current systems usually provide greater torque but they often require more maintenance for the motors. The use of motors generates dust and spark that can create hazards to the process. DC systems are common for the hobby robotics world as those systems are usually mobile, battery-powered robots. AC powered robots are common in industries and these often use Servo motors. Stepper motors are also used for these systems.

Servo Motor

A servo motor is a type of motor that can rotate with great precision. Normally this type of motor consists of a control circuit that provides feedback on the current position of the motor shaft; this feedback allows the servo motors to rotate with great precision. If you want to rotate an object at some specific angles or distance, then you use a servo motor. It is just made up of a simple motor which runs through a servo mechanism. If motor is powered by a DC power supply then it is called DC servo motor, and if it is AC-powered motor then it is called AC servo motor.

The servo motor is an assembly of four things: a normal DC motor, a gear reduction unit, a position-sensing device, and a control circuit. The DC motor is connected with a gear mechanism that provides feedback to a position sensor which is mostly a potentiometer. From the gearbox, the output of the motor is delivered via servo spline to the servo arm. For standard servo motors, the gear is normally made up of plastic whereas, for high power servos, the gear is made up of metal.

A servo motor consists of three wires- a black wire connected to the ground, a white/yellow wire connected to the control unit, and a red wire connected to the power supply.

The function of the servo motor is to receive a control signal that represents a desired output position of the servo shaft and apply power to its DC motor until its shaft turns to that position. It uses the position-sensing device to figure out the rotational position of the shaft, so it knows which way the motor must turn to move the shaft to the instructed position. The shaft commonly does not rotate freely around similar to a DC motor, however rather can just turn 200 degrees.

Servo Motor.

From the position of the rotor, a rotating magnetic field is created to efficiently generate toque. Current flows in the winding to create a rotating magnetic field. The shaft transmits the motor output power. The load is driven through the transfer mechanism. High-function rare earth or other permanent magnet is positioned externally to the shaft. The optical encoder always watches the number of rotations and the position of the shaft.

Stepper Motor

A stepper motor is an electromechanical device it converts electrical power into mechanical power. Also, it is a brushless, synchronous electric motor that can divide a full rotation into an expansive number of steps. The motor's position can be controlled accurately without any feedback mechanism, as long as the motor is carefully sized to the application. Stepper motors are similar to switched reluctance motors. The stepper motor uses the theory of operation for magnets to make the motor shaft turn a precise distance when a pulse of electricity is provided. The stator has eight poles, and the rotor has six poles. The rotor will require 24 pulses of electricity to move the 24 steps to make one complete revolution. Another way to say this is that the rotor will move precisely 15° for each pulse of electricity that the motor receives.

Types of Stepper Motor

There are three main types of stepper motors, they are:

- Permanent magnet stepper,
- Hybrid synchronous stepper,
- Variable reluctance stepper.

Permanent Magnet Stepper Motor

Permanent magnet motors use a permanent magnet (PM) in the rotor and operate on the attraction or repulsion between the rotor PM and the stator electromagnets. This is the most common type of stepper motor as compared with different types of stepper motors available in the market. This motor includes permanent magnets in the construction of the motor. This kind of motor is also known as tin-can/can-stack motor. The main benefit of this stepper motor is less manufacturing cost. For every revolution, it has 48-24 steps.

Variable Reluctance Stepper Motor

Variable reluctance (VR) motors have a plain iron rotor and operate based on the principle that minimum reluctance occurs with minimum gap, hence the rotor points are attracted toward the stator magnet poles. The stepper motor like variable reluctance is the basic type of motor and it is used for the past many years. As the name suggests, the rotor's angular position mainly depends on the magnetic circuit's reluctance that can be formed among the teeth of the stator as well as a rotor.

Hybrid Synchronous Stepper Motor

Hybrid stepper motors are named because they use a combination of permanent magnet (PM) and variable reluctance (VR) techniques to achieve maximum power in small package sizes.

The most popular type of motor is the hybrid stepper motor because it gives a good performance as compared with a permanent magnet rotor in terms of speed, step resolution, and holding torque. But, this type of stepper motor is expensive as compared with permanent magnet stepper motors. This motor combines the features of both the permanent magnet and variable reluctance stepper motors. These motors are used where less stepping angle is required like 1.5, 1.8 & 2.5 degrees.

Difference between Stepper Motor and Servo Motor

Servo motors are suitable for high torque & speed applications whereas the stepper motor is less expensive so they are used where the high holding torque, acceleration with low-to-medium, the open otherwise closed-loop operation flexibility is required. The difference between the stepper motor and servo motor includes the following:

Stepper Motor	Servo Motor
The motor which moves in discrete steps is known as the stepper motor.	A servo motor is one kind of closed-loop motor that is connected to an encoder to provide speed feedback & position.
Stepper motor is used where controls, as well as precision, are main priorities.	Servo motor is used where the speed is the main priority.

The overall pole count of the stepper motor ranges from 50 to 100.	The overall pole count of servo motor ranges from 4 to 12.
In a closed-loop system, these motors move with a consistent pulse.	These motors need an encoder to change pulses to control the position.
Torque is high in less speed.	Torque is low in high speed.
Positioning time is faster throughout short strokes.	Positioning time is faster throughout long strokes.
High-tolerance movement of inertia.	Low-tolerance movement of inertia.
This motor is suitable for low rigidity mechanisms like pulley and belt.	Not suitable for less-rigidity mechanism.
Responsiveness is high.	Responsiveness is low.
These are used for fluctuating loads.	These are not used for fluctuating loads.
The adjustment of gain/tuning is not required.	The adjustment of gain/tuning is required.

Hydraulic Robots

A robot that is powered by hydraulic actuators, usually controlled by servo valves and analog resolvers. Hydraulics is sometimes looked at as an alternative to electric motors. Some of the primary reasons for this include:

- Linear motion.
- Very high torque applications.
- Small package for a given torque.
- Large number of motors that can share the reservoir/pump can increase volume efficiency.
- You can add dampening for shock absorption.

However there are also some downsides to using hydraulics including:

- More parts are required (however they can be separated from the robot in some applications).
- Less precise control (unless you use a proportional valve).
- Hydraulic fluid (mess, leaks, mess, and more mess).

Hydraulic systems use an incompressible liquid (as opposed to pneumatics that use a compressible gas) to transfer force from one place to another. Since the hydraulic system will be a closed system (ignore relief valves for now) when you apply a force to one end of the system that force is transferred to another part of that system. By manipulating the volume of fluid in different parts of the system you can change the forces in different parts of the system. So here are some of the basic components used (or needed) to develop a hydraulic system.

Basic Hydraulic Components

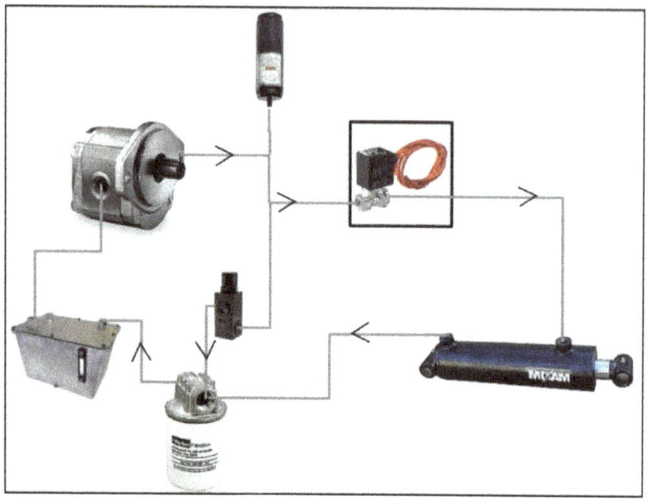

From the Reservoir the fluid goes to the Pump where there are three connections. 1. Accumulator(top) 2. Relief Valve(bottom) and 3. Control Valve. The Control Valve goes to the Cylinder which returns to a filter and then back to the Reservoir.

Pump

The pump is the heart of your hydraulic system. The pump controls the flow and pressure of the hydraulic fluid in your system that is used for moving the actuators. The size and speed of the pump determines the flow rate and the load at the actuator determines the pressure. For those familiar with electric motors the pressure in the system is like the voltage, and the flow rate is like the electrical current.

Pump Motor

We know what the pump is, but you need a way to "power" the pump so that it can pump the hydraulic fluid. Generally the way you power the pump is by connecting it to an electric motor or gas/diesel engine.

Hydraulic Fluid

Continuing the analogy where the pump is the heart, the hydraulic fluid is the blood of the system. The fluid is what is used to transfer the pressure from the pump to the motor.

Hydraulic Hoses and Fittings to Connect Things

These are the arteries and veins of the system that allows for the transfer of hydraulic fluid.

Hydraulic Actuators – Motor/Cylinder

Motor

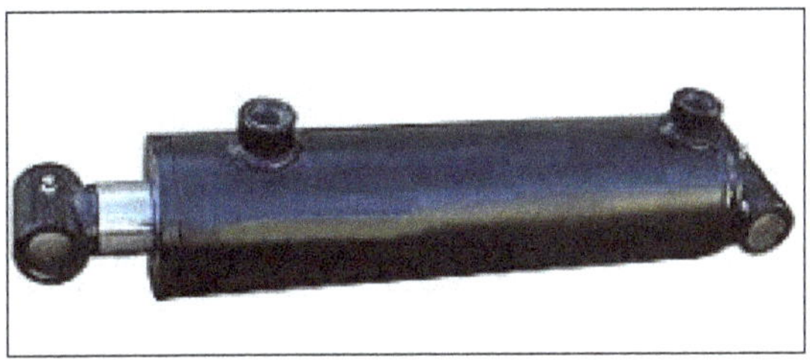

Cylinder

The actuator is generally the reason we are designing this hydraulic system. The motor is essentially the same as the pump; however instead of going from a mechanical input to generating the pressure, the motor converts the pressure into mechanical motion.

Actuators can come in the form of linear motion (referred to as a hydraulic cylinder) or rotary motion motors.

Cylinders you generally apply a force and the cylinder end extends, and then if you release the force the cylinder get pushed back in (think of a car lift). This is the classic and most common use of hydraulics.

For rotary motors there are generally 3 connections on the motor:

- A – Hydraulic fluid input/output line.
- B – Hydraulic fluid input/output line.
- Drain – Hydraulic fluid output line (generally only on motors, not cylinders).

Depending on the motor you can either only use A as the fluid input and B as the fluid output and the motor only spins in one direction. Or some motors can spin in either direction based on if A or B is used as the input or output of the hydraulic fluid.

The drain line is used so when the system is turned off the fluid has a way to get out of the motor (to deal with internal leakage and to not blow out seals). In some motors the drain line is connected to one of the A or B lines. Also there are sometimes multiple drain lines so that you can route the hydraulic hoses from different locations.

There are some actuators that are designed to be leakless and hold the fluid and pressure (using valves) so that the force from the actuator is held even without the pump. For example these are used in things like automobile carrying trucks that need to stack cars for transport.

Reservoir

This is essentially a bucket that holds the fluid. They are usually a little fancier so that they have over pressure relief valves, lids, filters, etc.

The reservoir is also often a place where the hydraulic fluid can cool down if it is getting hot within the system. As the fluid gets hotter it can get thinner which can result on increased wear of your motor and pump.

Filter

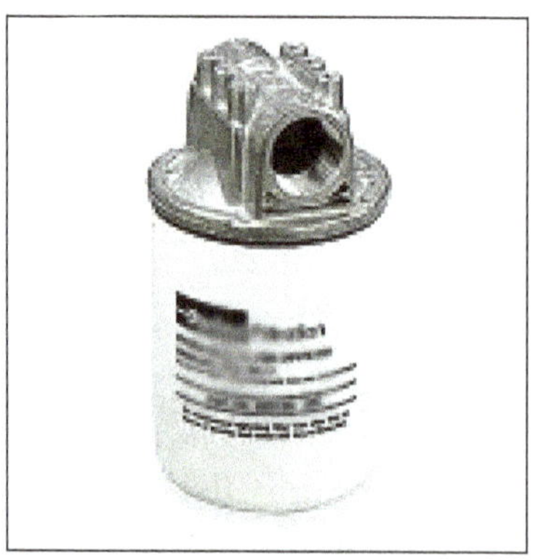

Keeps your hydraulic fluid clean before going back to the reservoir. Kind of like a person's kidneys.

Valves and Solenoids

Valves are things that open and close to allow the control of fluid. These can be controlled by hand (i.e. manual), or more often my some other means.

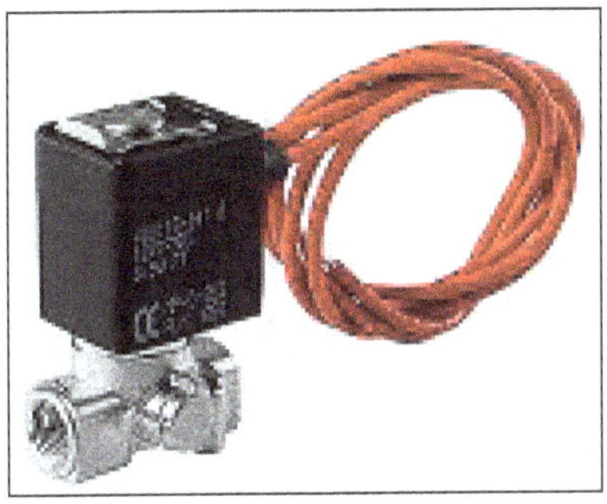

Valve (metal) with Solenoid (black) attached on top.

One common method is to use a solenoid which is a device that when you apply a voltage can be used to open a valve. Some solenoids are latching, which means you quickly apply a voltage and it opens the valves, and then you apply a voltage again (usually switching polarity) to close the valve.

Check Valves (One Way Valve)

These are a type of valve that can be inline to allow the flow of hydraulic fluid in only one direction.

Relief Valve

These are a type of valve that automatically opens (And lets fluid out) when the pressure gets too high. This is a safety feature so you don't damage other components and/or cause an explosion.

Pilot Valve

These are another special class of valve that can use a small pressure to control a much larger pressure valve.

Pressure and Flow-Rate Sensors/Gauges

You need to have sensors (with a gauge or computer output) to measure the pressure and/or flow-rate so you know how the system is operating and if it is operating how you expect it to operate.

Accumulator

The accumulator is essentially just a tank that holds fluid under pressure that has its own pressure source. This is used to help smooth out the pressure and take any sudden loads from the motor by having this pressure reserve. This is almost like how capacitors are used in electrical power circuits. The pressure source in the accumulator is often a weight, springs, or a gas. There will often be a check valve to make sure the fluid in the accumulator does not go back to the pump.

Advantages of Hydraulic Robot Arms

- Work Faster: To begin with, industrial or hydraulic robotic arms are faster in movement than human arms, but can be if handled without adhering to proper safety measures. However, they are often fitted with safety measures in the form of sensors to ensure humans enjoy some level of safety when handling or working with them. Many companies and factories use them to speed up the process of manufacturing, with precision and a high degree of accuracy.
- Reduce Costs: Working with several humans can be costly because a company has to part with a lot of money in labor and related costs. However, the robotic arm can do the work of humans at a much faster rate and at a more affordable cost. While it can be capital intensive because of the initial capital outlay required, the company saves money in the long run because maintenance costs are relatively lower than labor costs. Once the equipment is purchased, small costs are required for maintenance and minor repairs.
- Minimise Errors: Unlike humans who get tired and become inefficient, the robotic arm can work without the feelings of fatigue experienced by humans. Fatigue can take a toll on human workers or labour force and result in serious errors. However, working with the hydraulic robot arm helps to boost accuracy and minimise or eliminate errors.

Pneumatic Robots

Pneumatic robots are controlled with compressed air. They have advantages and disadvantages.

Fluid Movements

One of the perks of pneumatic robots is that they can move quite smoothly. When requiring lifelike robots for films or theme park attractions, the Walt Disney Company often depends on pneumatic robots to provide lifelike experiences for viewers and visitors.

Disney's experts came up with a pneumatic tube system for characters such as those in its Hall of Presidents attraction, with a transmission stored in a robot's body instead of its arms. As a result, the movements it makes are fluid and seem real compared to prior non-pneumatic robotics applications that required large, heavy motors.

Comparatively Low Safety Risk

In contrast to gas-powered robots that pose explosion and fire risks if they leak, the safety hazards associated with pneumatic robots are low because only air escapes into the environment if there's a problem.

This advantage is one reason some factories use pneumatic robots in automation applications to speed up the workflow. At its facility in Kansas, Simmons Pet Food depends on an automatic pneumatic system to direct product ingredients through the proper stainless steel channels on the production line.

The plant improved its productivity by 35 percent after expanding its previous pneumatic system. It can now keep facility operations running 24 hours a day.

Low-Maintenance Options

Provided the professionals using them make sure the air compressor components are of good quality before investing in a pneumatic robot, these solutions require less maintenance than some other kinds of robots. That characteristic makes them ideal for robots that have lots of moving parts that could wear out. The YuMi is an industrial machine that can work alongside humans as a collaborative robot or cobot. It has pneumatic grippers capable of handling delicate items, such as produce, as well as a wide assortment of other products.

Most similar grippers include hinges, increasing the possibility that they could wear out. However, the YuMi's pneumatic gripping system provides the movement with an air pressure and vacuum system, reducing the need for substantial ongoing maintenance. The system's grippers can even interlock, making them similar to human hands.

Less Control of Speed

When a company needs a robotic application that allows for precise tweaks of speed, an electric or liquid-powered system may be preferable to one controlled with air. That's because pneumatic systems do not allow for as much speed control compared to other options. Therefore, additional investments may be required to supplement a pneumatic robotics system with other parts that provide for more speed control.

Higher Long-Term Operating Costs than Hydraulics

Pneumatic systems have lower upfront costs compared to hydraulic systems, and that fact makes them initially appealing to businesses trying to keep costs low. However, over time, they could cost five to ten times more to operate than hydraulics.

Although a robot with hydraulic parts will be more expensive than a pneumatic one, it's also more efficient. That's because hydraulic systems work at higher pressures, making them compatible with small-scale actuators.

Prone to Moisture Buildup

If not cared for properly, pneumatic systems can have short lifespans due to a hidden problem: Condensation. When moisture builds up inside pneumatic pipes, it can compromise the functionality of the robot and affect how long the equipment lasts. Factors including the humidity and pressure of the supplied air can influence the probability of trapped moisture.

To reduce the occurrence, some facilities use dryers to dehumidify the air before it enters the pneumatic system. However, even that precautionary step is not always successful, especially with the use of small actuators.

Also, condensation increases as the diameter of a pipe increases and the pressure used in the application goes up. Therefore, experts recommend using the smallest pipes and lowest pressure possible. A pipe material with high thermal conductivity, such as copper, can also reduce the amount of moisture in a pneumatic system. Like anything else, pneumatic robots have associated pros and cons. The case studies and other specifics above indicate how essential it is to evaluate the task at hand and determine whether the process is best carried out with a pneumatic system or one that produces movement with a substance other than air.

Pneumatics are relatively easy to design and put together, plus they're lightweight, low-maintenance and quite powerful. A pneumatic system can, however, be more expensive, take up more space and require more programming. While pneumatic power can lead to more fluid and precise movement, but you don't have as much flexibility in the speed of movement.

How Pneumatics Work?

Before you start building a pneumatic robot, it can be helpful to have a basic understanding of how pneumatics works. A pneumatic system controls the movement of compressed air, like an electronic system control the flow of electrons. This air movement controls the movement of the robotic part.

A pneumatic system needs five main parts to work:

- The compressor, which produces compressed air using electricity, gasoline or propane.
- The reservoir, which stores the compressed air.
- The valves, which manage the air flow.
- The circuits, which control the valves.
- The actuator, which uses the air to accomplish the desired task.

What you'll Need?

Here are the basic supplies you'll need for your pneumatic robot:

- A compressor.
- A solid state relay (SSR), which you'll use to control the compressor, this needs to be rated for the maximum current of the compressor.
- A pressure relief valve, which connects to the compressor and releases air if the pressure gets too high.
- A pressure switch, which senses the air pressure and allows your software to turn off the compressor when it passes a certain threshold.
- Air tanks, which will act as reservoirs for the compressed air.
- Regulators, which limit the pressure sent to the pneumatic cylinders.
- Gauges, which tell you the air pressure in the system.
- Solenoid valves (electrically controlled valves).
- Push-to-connect fittings, which you'll use to connect all of the components together.
- Tubing, through which the air will flow.
- Pneumatic cylinders, which are the components that the pneumatic system controls to accomplish the task (actuators).
- A suitable microcontroller or control system.

- I/O board that can drive the relay and solenoid valves and read the sensors. An open collector driver would work well for the SSR and valves.

Nuclear-Powered Robots

Nuclear-powered robots used their own nuclear reactor that is smaller than the nuclear reactors of nuclear power plants or submarines. Nuclear powered robots are used by space agencies such as NASA for deep space exploration. Nuclear powered robots run for years and decades without the need for human interaction which makes them perfect fit for the space missions. However, if these robots are used on earth, there will be the need for proper disposal of nuclear material after the fuel is completely spent.

Green Powered Robots

We don't usually think of robots as being 'environmentally-friendly'. Until now, the closest thing we have seen to an earth-loving robot was is probably 'Wall-e'. But, the use of robots for sustainability can soon become a reality, thanks to green robots. Climate change needs to be checked more urgently than ever before, and we have already started leveraging technologies like artificial intelligence, blockchain, and IoT to fight climate change. Green robots, too, can support our fight against climate change. A variety of robots are already being developed that can help preserve the environment in various ways, from fighting wildfires, helping manage our waste and to cleaning our oceans and other water bodies. Here are a few ways in which green robots are helping with environmental sustainability:

Applications of Green Robots for Sustainability

The global temperature has risen to its highest point in the last 4000 years due to human activities. This can lead us on a path of self-destruction, if we don't act quickly. The availability of advanced technologies can help simplify our sustenance efforts and help slow down the effects of climate change. Green robots can help automate and carry out environment-friendly tasks with utmost efficiency, without tiring.

Fighting Wildfires

The United States alone witnesses about an average of 100,000 wildfires in a year. And wildfires are becoming a common occurrence all over the world in recent years. The destruction that wildfire brings to plant and animal species is well-known. Its effect on human health and infrastructure, too, is well documented. Robots can help in our efforts to control wildfires quickly and help avoid risking the lives of firefighters. Robots armed with fire extinguishers and water propelling agents can be leveraged to contain wildfires.

These robots can be equipped with GPS technologies, computer vision, heat sensors, and artificial intelligence technologies. These technologies can enable to control the robot remotely and can efficiently detect and extinguish fires. These robots can be used in areas highly inaccessible to humans or environments that pose a high risk to human lives. Robots having robotic arms and camera sensors can be used in search and rescue operations and help save and rescue animals caught in wildfires. Thus, robots can be used effectively to replace humans in high-risk operations associated with wildfires. They not only can carry out human procedures efficiently but also help save precious human lives.

Waste Management

Humans are consuming resources and are producing large quantities of waste, which needs to be managed and disposed of in an environmentally friendly manner. Simple robots are already being used for carrying out basic chores such as vacuuming, cleaning, and mowing and sorting garbage according to type. Household waste collection can be carried out by GPS and ai-enabled garbage collection robots. These robots can also collect data from sensors attached, and AI algorithms can determine the optimal garbage collection route and process. Thus, the process of garbage collection is streamlined, resulting in minimum use of resources, benefiting the environment. Advanced use of robots for waste management includes their use for waste treatment and recycling purposes. Automating the processes involved in waste management and waste treatment can help reduce CO2 emissions associated with these procedures. Robots can help reduce waste that is incinerated by efficiently sorting materials that can be recycled and put to use again. They can even help sort waste materials quickly and more efficiently than humans reducing the input power and costs associated with such processes.

Restoring Ocean Health

Human activities on land, industrial effluents, and water, such as oil drilling, have caused significant damage to our oceans. Out of these, oil and gas activities are a major cause of water pollution, and this was evident from the BP oil spill in 2010. Robots can be used to contain the harmful effects of such spills. A robot already has been designed that ingests microbes in the ocean and converts it into energy. The robot can operate continuously for months and can generate more energy than it consumes. Such a robot can not only help reduce the impact of oil leakages from tankers and ships but can also help remove chemicals from rivers and seas. The only current limitation of such robots is that they aren't made of biodegradable materials. Efforts are made to design robots of completely biodegradable materials so that they don't have any negative impact on the environment. Then, they can be deployed in thousands to clean every water body on the planet.

Robots can also be used for waste collection from our water bodies. Plastic waste can be collected by robots floating on the water bodies or can even be deployed on the ocean floor to collect waste from the depths of the ocean. Robots can prove to be highly cost and time-efficient in waste collection when compared to humans. Hence, their deployment will only gain prominence in the upcoming years.

Renewable Energy Sector

The adoption of renewable energy sources like solar and wind energy is gaining tremendous momentum. Robots can help simplify the processes involved in renewable energy generation, especially in solar energy sources. The power output of solar cells depends upon the solar energy that reaches the surface of the solar cells. The environment is one of the important factors that determine whether the cells will work with maximum yield. Dust can gather on the solar cells that can result in less output. The dust needs to be cleaned continuously, especially for solar farms located in high dust-density regions such as Africa or South Asia. Robots can automate the task of cleaning the solar panels whenever the dust layer over the cells passes permissible levels.

This ensures that the solar panels can provide maximum output and generate high amounts of energy. Another beneficial application of robots in the solar energy sector includes a robot with built-in solar panels. This robot can generate its own electricity while carrying out tasks assigned such as monitoring or cleaning and doesn't need to be dependent on an external source of energy.

Sustainable Agriculture

Robots can make agriculture more-environmentally friendly. They can be used to detect chemical levels in the soil. By analyzing the data, the optimum levels of fertilizers, whether chemical or natural, can be determined for maximum productivity. Other robots can be used to automate tasks such as planting, sowing, and watering. All these tasks can be carried out with precision and in a short time with the help of robots. Robots can also be used for crop harvesting, as robots attached with sensors and computer vision can determine the optimum time for picking up fruits and vegetables. Robots can also be used for livestock farming, for instance, where they can be used for milking cows or managing a herd of domesticated animals.

Classification based on Workspace Geometry

Cartesian Robots

Cartesian system is one that moves in three, orthogonal axes — X, Y, and Z — according to the Cartesian coordinates. (Although it should be noted that a rotary axis — in the form of an end effector or end of arm tooling — is sometimes included on the outermost axis of a Cartesian robot).

What makes a Cartesian robot a robot is that the axes perform coordinated motion, through a common motion controller.

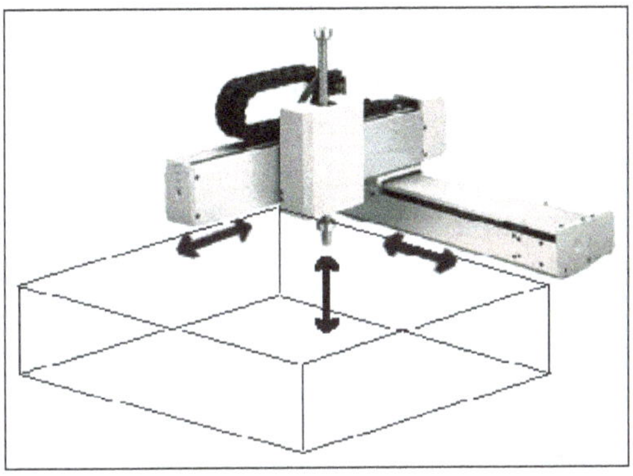

The axes of a Cartesian robot are made from some form of linear actuator – either purchased as a pre-assembled system from a manufacturer or custom-built by the OEM or end user from linear guide and drive components. An automatically controlled, reprogrammable, multipurpose manipulator programmable in three or more axes, which can be either, fixed in place or mobile for use in industrial automation applications.

But not every linear system that works in the X-Y, or X-Y-Z axes is a Cartesian robot. One notable exception is a type of robot that employs two base (X) axes in parallel. This configuration – 2X-Y or 2X-Y-Z, for example – moves the robot out of the Cartesian category and into the category of gantry robots.

The primary difference between gantry and Cartesian robots is that a Cartesian robot uses one linear actuator on each axis, whereas a gantry robot is always constructed with two base (X) axes, with the second (Y) axis spanning them. This configuration prevents the second axis from being cantilevered and allows gantries to have much longer stroke lengths – and in many cases, larger payloads – than Cartesian robots.

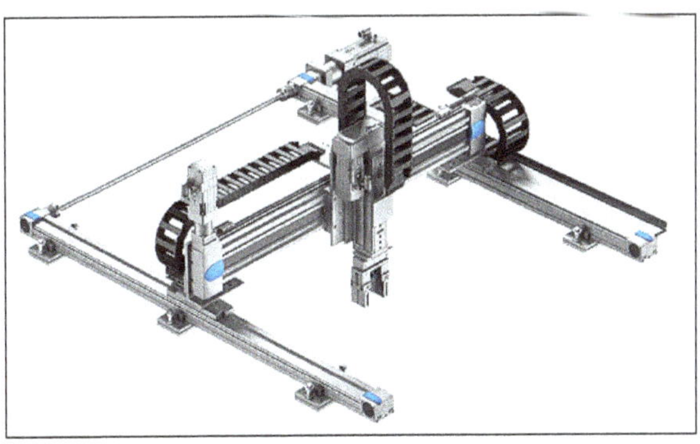

A gantry configuration can include two Y and/or two Z axes, for additional load capacity and stiffness, but the defining feature of a gantry robot is its two base (X) axes.

The second type of multi-axis linear system that does not fall under the definition of Cartesian robot is the XY table. The difference between Cartesian robots and XY tables lies in the mounting and loading arrangement. In a Cartesian robot, the second or third (Y or Z) axis is cantilevered, being supported at only one end by the axis below it. In addition, the load on the outer axis is generally cantilevered from that axis.

This arrangement creates not only a moment load on the outer axis, due to the applied load, but also a significant moment load on the supporting axis, due to the combined effect of the applied load along with the outer axis. The mounting and loading arrangement limits the load-carrying capability of Cartesian robots and is a primary factor in determining the maximum stroke length for the outer (cantilevered) axis.

In contrast, XY tables consist of two axes centered on top of each other, often with similar stroke lengths. In addition, the load is generally centered on the Y axis. This axis configuration and load positioning results in very little cantilevered loading on either axis (and often no cantilevered loading on the Y axis).

Unlike Cartesian systems, which have significant cantilevered loads, XY tables have very little cantilevered loading on either axis.

Cartesian robots overlap SCARA and 6-axis (articulated) robots in some technical specifications and can be applied in some of the same applications, but Cartesian robots have several benefits over SCARA and 6-axis types. First, Cartesian designs provide a rectangular work envelope in which a significant percentage of the robot's footprint is used as active work area. SCARA and 6-axis types, on the other hand, have circular or oval work envelopes that often result in a lot of dead (unused) space, especially when the required travel, or reach, is very long.

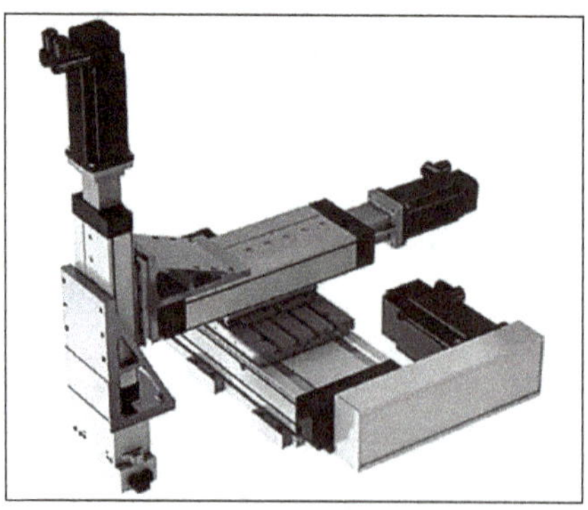

The base (X) axis of a Cartesian robot is, in most cases, supported along its entire length, but the Y and/or Z axis is cantilevered.

Cartesian robots can be constructed from virtually any type of linear actuator with any variety of drive mechanisms — belt, ball or lead screw, pneumatic actuator, or linear motor. (Note that rack and pinion drives are also possible, but are more commonly used in gantry systems with very long strokes.) This means they can, and often do, have better positioning accuracy and repeatability than SCARA and 6-axis types. Cartesian robots also have an ease-of-use advantage in terms of programming because their kinematics is simpler (three Cartesian axes, rather than multiple rotational axes).

In the recent past, pre-assembled Cartesian robots were rare, with most units being custom-built by an OEM, a robot integrator, or even the end user. But now, many linear actuator manufacturers also provide pre-configured, pre-assembled Cartesian systems, with myriad options to fit common travel, payload, speed, and precision requirements. And manufacturers of traditional 6-axis and SCARA robots are getting in on the action as well, recognizing that for many industrial automation and assembly applications, Cartesian robots offer a better tradeoff between load capacity and footprint than SCARA and 6-axis designs.

Cylindrical Robots

Cylindrical robots have a primary arm that moves up and down. Built within the robotic arm is a cylinder that creates this motion by extending and retracting itself. In many of these cylindrical robotic models, the rotation is generated by gears and a motor, and the vertical motion is generated using a pneumatic cylinder.

Cylindrical robots use a coordinate system with three dimensions that use a selected reference axis and relative distances from it for specifying point positions. Point positions are also specified by using the distance to a selected reference direction and the axis' relative direction, and by the distance to the axis perpendicular from a selected reference plane.

These types of robots are useful for objects that need to have rotational symmetry around their longitudinal axes, such as a metal cylinder for heat distribution, long, straight wires that produce electromagnetic fields, and the rounded cross-sections of straight pipes.

Some applications of cylindrical robots include:

- Spot welding,
- Die-cast machine handling,
- General machine handling tools,
- Grinding processes,
- Palletizing,
- Assembly operations.

Spherical Robots

A spherical robot is a robot with two rotary joints and one prismatic joint; in other words, two rotary axes and one linear axis. Spherical robots have an arm which forms a spherical coordinate system.

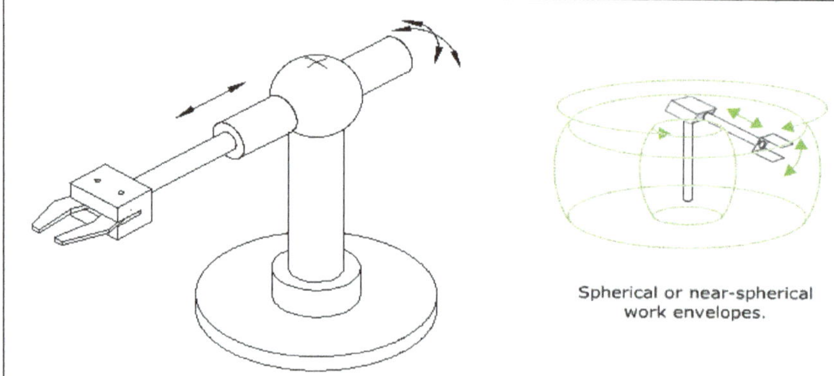
Spherical or near-spherical work envelopes.
The Spherical Robot has three or more axes of movement.

Spherical Coordinate System

In mathematics, a spherical coordinate system is a coordinate system for three-dimensional space where the position of a point is specified by three numbers: the radial distance of that point from a fixed origin, its polar angle measured from a fixed zenith direction, and the azimuth angle of its orthogonal projection on a reference plane that passes through the origin and is orthogonal to the zenith, measured from a fixed reference direction on that plane. The radial distance is also called the radius or radial coordinate. The polar angle may be called colatitude, zenith angle, normal angle, or inclination angle.

Articulated Robots

Articulated robots are the iconic mechanical denizens of the futuristic factory-floor. For decades, this image has seeped its way into our collective cultural consciousness and pervaded everything from cartoons to movies and even stock images. Also, though we've moved beyond the basic articulated design to some stunning, cutting edge machines, the image is so ingrained in our memories that the industrial robotics manufacturers continue to produce them to this day. But image alone isn't enough to keep this reliable, not-yet-antiquated, style of robot around. What are the benefits provided by articulated robots? Why do so many factories choose to rely on this classic, iconic workhorse?

What's an Articulated Robot?

Articulated robots are the most common types of industrial robots. Their resemblance to a human arm is perhaps one of the reasons they stand out so much in our minds. However, the mechanical benefits of this design are the real reason for their ubiquity and longevity. The arm design combines an extensive range of rotation motion and linear reach with the advantages of precision movement. Articulated arms are ideal for welding, material handling, pick-and-place operations, and dispensing. Their numerous axes and degrees of freedom mean that there is virtually no point in their work envelope that they cannot reach. This makes articulated robots one of the most versatile, flexible, and compact designs on the market.

Articulation of the Robot

The secret to the motion of articulated robots is their rotary joints. Joints can range from a two-axis configuration to something as complicated as ten or more axes. Articulated robots with six-degrees of freedom are the most common. Servo motors powering the joints act like bezel gears, changing the direction of motion by up to ninety-degrees while offering dial-in stops for precision programming and movement.

Work Envelope of an Articulated Robot

With any industrial robot, the work envelope is a significant factor in assessing its usefulness. Articulated arm robots can use a majority of their work envelope, which is one of their greatest advantages. The only part of the envelope they can't use is the back where the cables are located. However, some modern designs feature internally routed power and data cables that eliminate this problem and let the articulated arm take advantage of its entire sphere of reach.

Regardless of how the cables are routed, even the most basic articulated robot can maximize the usable space for its footprint on the factory floor. This is a major advantage to factories that have to consider production flow, safety, and floor space.

Speed of an Articulated Robot

One of the drawbacks of articulated robots is their speed. They are not as efficient as other types of robots that can accomplish very high-speed tasks. Articulated robots, due to their numerous joints and degrees of freedom, require complex kinematics to govern their motion. They also have higher component mass, so there is an inertial barrier to overcome at every change in direction. If speed is a crucial component of a factory's cost-benefit-analysis, then articulated robots may not be the ideal choice.

So, what will an articulated robot cost you? As with any factory upgrade, there are a few ways to answer this question. First, there is the upfront cost: what is the price of the actual unit? Then, there are operational and maintenance fees—these address the costs to operate and maintain the machines every month. However, if you also look at where you'll see savings, you'll get a better overall picture of your return on investment. Over time, the benefits of an articulated robot offset the upfront cost. For instance, factory automation improves the health and safety of workers by minimizing repetitive stress and limiting time in hazardous zones. Additionally, your company will produce products of more consistent quality. The result is increased productivity, which means that your investment will pay for itself down the road.

With so many factors constituting the "real cost" of robot upgrades, it's best to consult the experts on DIY-Robotics.com for a full cost breakdown. In the meantime, you can build your own miniature/prototype robot and familiarize yourself with the mechanics and programming. If you'd like to boost your factory's productivity with a robot that is flexible, agile, and has a sizable work envelope, you should consider an articulated robotic arm.

Delta Robots

Delta robots are robots with a base connected to jointed parallelograms. These parallelograms perform motions in a solitary End of Arm Tooling (EOAT), within a workspace that is dome-shaped. This type of robot is well-known in the industrial field for its ability to execute minute, precise motions. Delta robots can be called other names in the industrial field, such as spider robots or parallel robots.

Delta robots are primarily operated when production is required for groups of objects and these objects must be placed in either an assembly pattern or inside of a container. With the addition of capable vision technologies, delta robots can properly identify and select a range of object colors, shape options and sizes. They can also use these vision technologies to pick and place working from a programmed pattern or design.

These types of robots can work with more than 100 parts per minute, picking; placing and sorting them much faster than manual labor processes are capable of performing. Their design allows them to very efficiently perform repetitive tasks at a highly consistent rate. A chief benefit of delta robots revolves around the fact that its relatively heavy motors are attached to the robot's frame, permitting the robot's moving parts to be very light by comparison.

Some applications of delta robots are:

- Packing industry,
- High-precision assembly operations,
- Medical/Pharmaceutical operations,
- Food processing,
- Cosmetics,
- Soldering,
- Adhesive dispensing.

Classification of Robots on the basis of Degrees of Freedom

Autonomous Robots

An automatic robot is a type of manipulated robotic system considered to be as one of the earliest robotic systems on the basis of the control system it possesses. Automatic robots are divided into four main categories based on their characteristics and applications.

Manipulation robotic system is classified into three types:

- Autonomous controlled robots,
- Remote controlled robots,
- Manually controlled robots.

Autonomous robots are mainly used in industrial areas; whereas the remote controlled robots are used in environments that are restricted for human beings. The manually controlled robots are widely used for handling goods and also for transportation.

Types of Autonomous Robotic System

Out of three types of manipulation robotic system, the autonomous system is further classified into four types:

Programmable Automatic Robots

A programmable robot is a first generation robot with an actuator facility on each joint. The robots can be reprogrammable based on the kind of application they are commissioned to. The function and application of the robots can be changed by reprogramming after the robot is programmed once to perform a function in the given pattern and fixed sequence.

Robot kits like Lego mind storms, Bioloid from programmable Robotics can help the students to learn about its programming and working. The advanced mobile robot, robotic arms and gadgeteer are some of the examples of these programmable robots.

The main drawback of this autonomous robot is that once programmed it persists operation even if there is a need to change its task (in the event of emergency). These robots can be used in different applications like mobile robotics, industrial controlling and space craft applications.

Non-Programmable Automatic Robots

This robot is one of the basic types of robot, in fact, a non-programmable robot. This robot is not even considered as a robot, but is an exploiter lacking reprogrammable controlling device. The mechanical arms used in industries are some of the examples of these types of robots wherein the robots are generally attached to the programmable devices used in industries for mass production as shown in the figure.

These types of robots find applications in some of the devices including path guiders and medical products' carriers and also some line follower robots.

Adaptive Robots

Adaptive robots are also industrial robots that can be adapted independently to various ranges in the process. However, these robots are more sophisticated than programmable robots. These can be adapted up to a certain extent, and after evaluation they can perform the action required in that adapted area. These robots are mostly equipped with sensors and control systems.

Sensors are used to sense environmental conditions, process variables and other parameters related to a particular task. Feedback control system accesses these signals from the sensors, and depending on the algorithm implemented, it controls the outputs.

Adaptive robots are mainly used in applications such as spraying and welding systems. Robotic gripper and 2- finger adaptive gripper are examples of this autonomous robot.

These robots can be used in different applications like aerospace, medical, consumer goods, house-hold applications and manufacturing industrial areas.

Intelligent Robots

Intelligent robots, as the name suggests, are the most intelligent of all the other types of robots with sensors and microprocessors for storing and processing the data. These robots performance is highly efficient due to their situation-based analyzing and task performing abilities. Intelligent robots can sense the senses like pain, smell and taste and are also capable of vision and hearing, and – in accordance, perform the actions and expressions like emotions, thinking and learning.

These robots find their applications in the fields like medical, military applications and home appliance control systems, etc.

Redundant Robots

Redundant robots are robots that can accomplish the same task in more than one way. Our arms are redundant. You can touch your nose and then raise your elbow. If there are obstacles you use the redundancy to avoid them. You use the redundancy to increase your comfort. A redundant manipulator has more than six degrees of freedom which means that it has additional joint parameters that allow the configuration of the robot to change while it holds its end-effector in a fixed position and orientation. A typical redundant manipulator has seven joints, for example three at the shoulder, one elbow joint and three at the wrist. This manipulator can move its elbow around a circle while it maintains a specific position and orientation of its end-effector. A snake robot has many more than six degrees of freedom and is often called hyper-redundant. These robots are capable of operating in tightly confined places. These robots can have infinite possible ways to place and orient the object.

Deficient Robots

Deficient robots have a degree of freedom less than six. Robots with three degrees of freedom are able to pick objects and place it anywhere on the work envelope using the three coordinate axes.

Classification of Robots based on their Application

Industrial Robots

Because they can be programmed to perform dangerous, dirty and/or repetitive tasks with consistent precision and accuracy, industrial robots are increasingly used in a variety of industries and applications. They come in a wide range of models with the reach distance, payload capacity and the number of axes of travel (up to six) of their jointed arm being the most common distinguishing characteristics.

In both production and handling applications, a robot utilizes an end effector or end of arm tooling (EOAT) attachment to hold and manipulate either the tool performing the process, or the piece upon which a process is being performed.

The robot's actions are directed by a combination of programming software and controls. Their automated functionality allows them to operate around the clock and on weekends—as well as with hazardous materials and in challenging environments—freeing personnel to perform other tasks. Robotic technology also increases productivity and profitability while eliminating labor-intensive activities that might cause physical strain or potential injury to workers.

Industrial robots are used in a variety of applications. These include:

- Handling: Capable of manipulating products as diverse as car doors to eggs, industrial robots are fast and powerful as well as dexterous and sensitive. Applications include pick and place from conveyor line to packaging, and machine tending, where raw materials are fed by the robot into processing equipment such as with injection molding machines, CNC mills and lathes and presses.
- Palletizing: Industrial robots load corrugated cartons or other packaged items onto a pallet in a defined pattern. Robotic palletizers rely on a fixed position or overhead gantry robot with special tooling that interfaces with the individual load components, building simple to complex layer patterns on top of a pallet that maximize the load's stability during transport. There are three primary types of palletizing: inline or layer forming, depalletizing or unloading, and mixed case.

- Cutting: Due to their dangerous nature, laser, plasma and water jet cutters are frequently used with robots. Hundreds of different cutting paths can be programmed into the robot, which produces precise accuracy and path following with greater flexibility than most dedicated cutting machines.
- Finishing: Multi-axis robots can grind, trim, fettle, polish and clean almost any part made in any material for a consistent quality finish.
- Sealing and Gluing: To apply sealant or glue, a robot follows a path accurately with good control over speed while maintaining a consistent bead of the adhesive substrate. Robots are frequently used for sealing applications in the car industry to seal in windows, as well as in packaging processes for automated sealing of corrugated cases of product.
- Spraying: Due to the volatile and hazardous nature of solvent-based paints and coatings, robots are used in spray applications to minimize human contact. Paint robots typically have thin arms because they don't carry much weight, but need maximum access and movement fluidity to mimic a human's application technique.
- Welding: Used for both seam (MIG, TIG, arc and laser) and spot welding, robots produce precise welds, as well as control parameters such as power, wire feed and gas flow.

How are Robots Used?

Robots are used in a variety of ways throughout manufacturing and distribution:

- Load Building: Assembling a pallet load of products at the end of a production line.
- Manufacturing: Performing processing and assembly functions to work-in-process.
- Quality Control: Testing and inspection procedures deploy robots for repetitive or dangerous work.
- Transportation: Loading pallets prior to shipping.
- Warehousing: Removing received products from pallets and routing them to storage locations within a facility.

What are the Benefits?

Industrial robots provide a variety of benefits:

- Accuracy: Robotic palletizers are software-directed for proper load placement.
- Flexibility: Robotic systems can be re-purposed for other uses; end effectors can be switched out to handle different load types.

- Lower Labor Costs: Automated pallet building reduces worker strain and frees operators for other tasks.
- Quiet Operation: Servo-based, robotic palletizers generate low noise levels.
- Reduced Product Damage: Gentle handling prevents package and product damage.
- Speed: The systems increase rate productivity up to 50%.

Where are Robots Used?

Industrial robots are used in many industries, including:

- Aerospace,
- Automotive,
- Beverage,
- Computers,
- Consumer goods,
- E-Commerce,
- Electronics,
- Food,
- Grocery,
- Hardware,
- Healthcare,
- Liquor distribution,
- Manufacturing,
- Medical products,
- Pharmaceutical,
- Quality control and inspection,
- Retail,
- Warehousing and distribution.

Non-Industrial Robots

We can divide nonindustrial robots into 4 groups:

- Mobile robots,

- Exoskeletons,
- Walking machines,
- Prostheses.

A mobile robot has a variety of applications nonindustrial robots type, they can differentiate by their drive force as wheel-type and crawler tracks.

Ways to control mobile robots are:

- Remotely controlled mobile robot by wires or radio.
- Programmed mobile robot or autonomously controlled.
- Guided by rails or other ways.

Mobile robots can operate in various situations like harmful and harsh environment, remote location; can handle dangerous and hazardous objects.

Exoskeletons are wearable devices that work together with a human; they act as an amplifier for human actions in environments where normally movements are hindered. Exoskeletons are made of rigid or soft materials depending on the application, they are also equipped with sensors and actuators, and they can cover the entire human body or only parts of body. Main purposes of exoskeleton use cases are:

- Amplify human muscular activity.
- Free human skeleton from unnecessary loads.
- Feedback about manipulated object measurements.

Exoskeletons are often has double layer structure, where internal layer is responsible for control mechanism and designed to interact with manipulator. The external layers follow internal layer movements and amplify them. Walking machines is a type of exoskeleton.

Prostheses are the combination of the walking machines and exoskeletons used by human who damaged or lost the limb. It is important for prostheses to be light-weight with short time between command and action. The easiest way to make this idea truth is to use bioelectricity and relevant muscles amplification.

References

- Servo-motor-working-and-basics: circuitdigest.com, Retrieved 12, February 2020
- Stepper-motor-types-advantages-applications: elprocus.com, Retrieved 05, June 2020
- Using-hydraulics-for-robots-introduction: robotsforroboticists.com, Retrieved 24 March 2020
- How-to-build-your-own-pneumatic-robot: workshopshed.com, Retrieved 03, March 2020

- How-green-robots-are-helping-with-environmental-sustainability: allerin.com, Retrieved 11, April 2020
- Different-types-of-autonomous-robots-and-real-time-applications: elprocus.com, Retrieved 05, july 2020
- What-is-a-cartesian-robot: linearmotiontips.com, Retrieved 25, june 2020

3

Basics of Robot Design

A robotic design involves the creation of a plan or convention for the construction of a robotic system. Science and technology have undergone rapid developments in the past decade which has resulted in the discovery of significant tools and techniques in the field of robotic design. A few of such tools and techniques are physical design, design software, tools and equipment etc. which have been extensively detailed in this chapter.

Physical Design and Design Software

- Identifying the purpose of a construction.
- Identifying specific requirements.

You are confronted with a situation. Here are two examples:

A community wants to construct a robot zoo in which the "animals" move their heads open their mouths and make appropriate sounds when they sense that someone is coming towards them. Design and build a prototype device which could satisfy this need.

A local pet shop wishes to sell a range of devices which automatically feed small cage pets (such as rabbits, gerbils, mice etc.) when their owners are away for the weekend. Design and build a prototype device which could satisfy this need.

You need to determine what problem you are trying to solve before you attempt to design and build a robot to solve a problem. Take the time to study a number of different situations and once you have decided what the situation is and you understand exactly what the problem is then write a design brief in a log book (this will be your working document as you work on your robot. This log book can be a paper notebook or an electronic document.) This is a short statement which explains the problem that is to be solved.

Researching and Designing

- Gathering information.
- Identifying specific details of the design which must be satisfied.
- Identifying possible and alternative design solutions.
- Planning and designing an appropriate structure which includes drawings.

Having written a brief, you are now ready to gather information which will help you to produce a successful design. First you will need to decide what information you require. This will be different from project to project and will also depend on the amount of information and knowledge you already have. A useful step will be to use the following chart. This will help you plan the type of information you will need to gather.

Gathering Information	
1. What is the practical function of the design? (What must my robot do?)	A design's practical functions can include: • Movement: How will the robot move within its environment? If it were put in a different environment, would it still be able to move within this new space? • Manipulation: How will the robot move or manipulate other objects within its environment? Can a single robot move or manipulate more than one kind of object? • Energy: How is the robot powered? Can it have more than one energy source? • Intelligence: How does the robot "think?" What does it mean to say that a robot "thinks?" • Sensing: How will my robot "know" or figure out what's in its environment? If it were put in a different environment, would it be able to figure out this new environment?
2. What part does appearance (shape and form, surface texture, colour, etc.) play in the design's function? What does the robot look like? Is there a reason for it to look as it does?	• Shape and form are important to a design's aesthetic qualities, ergonomics, strength, stability, rigidity, safety. • Surface texture, finish and colour can be appropriate to a design's: aesthetic qualities, mechanical, optical and thermal properties, durability, etc.
3. What materials are suitable for the design?	The properties of a material will determine its suitability for a design. For our work with robotics we have chosen to work with LegoT™. However, there are many different types of materials that can be and are used in the construction of robots. • Strength, hardness, toughness, density. • Durability. • The aesthetic qualities determined by colour, surface texture, pattern, etc. The materials cost and availability is also important factors.
4. What construction methods are appropriate to the design?	Construction techniques fall into the categories of: • Cutting and shaping. • Fabrication - the assembly of the parts using screws, bolts, glues, solder, etc. • Moulding by the application of a force on the material. • Casting - using a mould to form the shape of a solidifying material.

	A particular material can only be worked in a limited number of ways. The method of construction therefore will be determined by the chosen material, the availability of manufacturing facilities, the skills of the work force and the production costs.
5. What are the likely social and environmental effects of the design?	The manufacture, use and disposal of any product will have both beneficial and detrimental effects upon people, wildlife and the environment. The designer therefore, has an enormous responsibility to consider very carefully the potential effects of any new design. This will include: health and safety factors, noise, smell, pollution, etc.

Gathering information can involve reading, listening, conducting interviews and observing. A specification is a detailed description of the problem to be solved. It should 'spell out' exactly what the design must achieve.

Creating a Prototype

- Testing the design,
- Troubleshooting the design.

You should ideally think of at least three different ways to solve the problem before you concentrate on any one in particular. Sketches and notes are required at this stage. You can also create prototypes using lego for this step. Once you have created a lego prototype, take a digital picture of it. Print out the picture and jot your notes below the picture in your log book. Once you have settled on one solution, go back over the list of specifications you have made. Make sure that each specification is satisfied.

Now it the time to produce some working drawings. These are the drawings that will assist you as you begin constructing the prototype of your structure. (Here again, lego and a digital camera might be your best friend.) You may choose to do your drawings by hand or you might want to use a draw program on the computer to assist you.

Determine a working schedule for yourself. Draw up a timetable showing how much time you expect to spend on each part of the design process. Your planning should also ensure that you have all the necessary materials and equipment that you need to complete your project.

Building your Robot

Construction work can now begin:

- The Art of LEGO Design by Fred Martin - An excellent resource for building very strong structures.

Programming and Testing your Robot

Now it is time to program your robot. This can be achieved in many different ways. Use can achieve rudimentary intelligence in your robot by using only relays, potentiometers, bump switches and some discrete components. You can increase complexity in intelligence in your robot by adding more sensors and continuing in the same vein of using hardwired logic. By introducing a more sophisticated control element, the microprocessor, you introduce a significant new tool in solving the robot control problem. Once you have written your program and downloaded into the RCX brick using the Infrared Sender, it is time to test your robot to see if it truly does what you want it to do.

Evaluating your Robot

- Evaluate the design,
- Evaluate the planning process.

As building and programming work progresses, and the design begins to take shape, you will automatically carry out tests on the design. You will also need to complete systems tests at various stages of the construction. If any of the tests show that you have failure in a joint, or that part of your structure is not meeting specifications, then you will have to make modifications in your plan.

When building and programming is complete, the entire project must be tested to see if it does the job for which it was designed. An evaluation needs to then be written. This should be a statement outlining the strengths and weaknesses in your design. It should describe where you have succeeded and where you have failed to achieve the aims set out in the specifications.

Tools and Equipment

Robot Design Tools and Techniques: Mechanical

As a sculptor chooses his own tools, so a roboticist selects his own design platforms. "We use SolidWorks daily for solid modeling. Such mechanical CAD tools are essential for designing systems that are as complex as most robots," says Derek Scherer, Robot & Automation engineer, The Golem Group LLC, who designed robotics for the movies: "The Hobbit: An Unexpected Journey", "Man of Steel" and "Elysium".

SolidWorks brings a deep dimensional and mechanical understanding of surfaces to bear on robot projects. When a desktop engineer cuts a sphere in half in SolidWorks, this platform has the same realization that a human being has of the qualities and characteristics of half a sphere in a 3D space. "SolidWorks produces diagrams, tolerances—everything a shop would need to fabricate the resulting design," says Scherer.

Another tool, Linkage, is a mechanical simulator that enables the user to create links and their connections and move parts inside the simulation to see how these operate collectively. "It's useful in uncovering major design flaws early on," says Scherer.

RobotsLAB, makers of RobotsLAB BOX educational robots uses Autodesk Inventor to design robotic mechanisms. While the capabilities of this design tool are many, the designer must still factor in the limits of fabrication technologies.

"If you want to cut the part out of metal, you have to keep in mind whether the cutter bit can make holes only in one size or can drill a hole on the underside of the part," says Ryan Wood, lead engineer, RobotsLAB. Each building material will have its own fabrication limits.

Blender, a popular physics animation tool, is a popular choice for animation demonstrations and it's free. "Whenever you see a robot animated on YouTube or a company website, chances are that they are using Blender," says DJ Sures, president of EZ-Robot, maker of personal and educational robot platforms.

Blender is a 3D CAD design package that enables the engineer to load design files, animate any design, add effects and visualize robot projects. "Blender uses real physics with animations to simulate your robot so you can verify that it operates as expected," says Sures.

Component Tips

RobotsLAB uses Autodesk Inventor to model components that it doesn't build. When considering how to use a sensor or motor to purchase for their robots, designers will bring a model of that into CAD software to determine its placement and to consider how to place parts and equipment around it. Some companies provide CAD models of the parts they sell. "But we may have to use a mechanical drawing of a unit to model it in our CAD software. Or, we may purchase the part and create a model from that."

If a motor or other component takes up too much space on the robot, a roboticist might try to re-design the other components or the enclosure that everything fits into, or they may try to select a motor in a smaller size, according to Wood. But these component and design choices affect the robot and its capabilities, too.

Integrating Electronics

EZ-Robot, maker of personal and educational robot platforms, uses Altium Designer from Altium's line of PCB (printed circuit board) design products. "We've used some of the free PCB design tools but there aren't a lot of free tools that are very good because the paid versions include lots of design samples and templates that the free versions don't," says Sures.

Ben Wirz, president of Element Products, uses the schematic capture tools DxDesigner and the PADS PCB design tool, both from Mentor Graphics. A typical project using these tools starts with a specification to follow.

Ben Wirz used DxDesigner and PADS to design the S2 Scribbler Robot.

The design flow starts with a paper design, using spreadsheets, says Wirz. In fact, Microsoft Excel is one of Wirz's most valuable electronics design tools. "We use it to do all the design equations, to calculate the resistor values and capacitor values and to work through all the theoretical calculations," says Wirz.

Microsoft Excel enables Wirz to easily document the electronics design process as design choices change, such as can happen with the ambient operating temperature specification. "If a customer says: 'We told you we wanted to work at 30 °C, but we want to work at 20 °C', we can change the number in Excel and see how this affects the design," says Wirz.

Wirz does the theoretical design and the mathematical design equations for engineering the circuit and enters the circuit into a schematic capture tool to add the transistors, resistors, and captures. "We make the circuit in DxDesigner, which connects all the pins together," says Wirz.

The Mentor Graphics HyperLynx Analog add-on software that works with DxDesigner Schematic Capture Software comes in handy at this point for simulations. "It lets me do electrical simulations of the circuitry we design. We can simulate analog circuits with it. We find it useful when doing any sort of analog design or filter design," says Wirz.

"Then we export it into the PCB tool to turn the theoretical into the physical, designing the circuit board and determining where the chips and components go down," says Wirz.

Once the circuit board is designed, the desktop engineer will want to integrate the board with the mechanical design. "For that we export the PCB design to Creo, a mechanical 3D design package from PTC," says Wirz. Creo enables the engineer to have a physical object to integrate with the robotics design.

"We can make a 3D model of an enclosure and print it out on my 3D printer to check that my circuit boards fit."

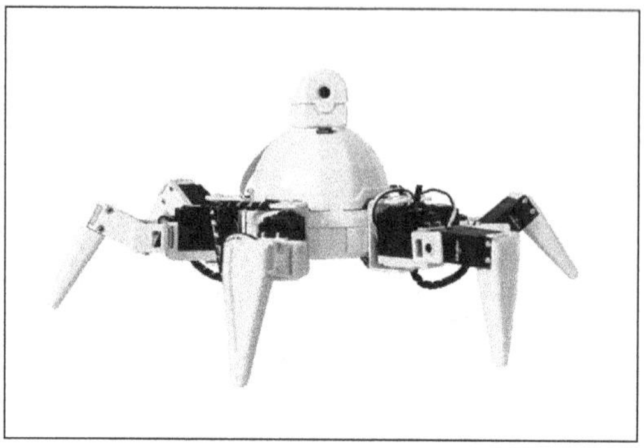

EZ Robot built a robot platform using Blender and Altium Designer.

Design Considerations

RobotsLAB designers consider the whole robot system, its electronic, computerized and mechanical aspects together and how these affect each other when designing within any of those areas.

"Say you need a robotic arm to pick up a 2-lb. weight. The gripper you design weighs a certain amount too. The base joint of the arm now needs to lift the weight as well as the weight of the arm and gripper. This leads us to question the power abilities of the electronics," says Wood.

More powerful electronics may be necessary if the performance of the robot requires more energy than the electrical specifications support, and to avoid power supply failure, for the safety of others in the robot's environment.

Similarly, to give the end effector more gripping strength, the roboticist may have to consider larger, more powerful gripper motors. Those motors could necessitate stronger motors in the rest of the arm to lift that extra weight, which could again lead to reconsidering which electronics could support the design. As each robot project is unique and fluid, so too is a roboticist's choice of tools.

Electronic Components

Electronics can at first seem extremely complicated to understand and learn. One look at a circuit board with all those little blinky LED's and black chips and unidentifiable circle pointy things can make anyone quit before starting.

But actually electronics can be much simpler than you think. Learning electronics is more like learning a foreign language alphabet. At first glance it is all a bunch of squiggles. But actually each letter has its own pronunciation and its own rules of use. And certain combinations of letters in a certain order form a word of some meaning. And a combination of words forms a sentence. This is the same for a circuit board. Each tiny component, such as a resistor or capacitor or transistor, has special rules and abilities. Combining a few into a circuit can create interesting effects. Combine a bunch of unrelated circuits together and suddenly you have a robot. So your first step would just to be to learn and understand the smallest of the components. Once there you can learn about combining them. Just like learning a foreign alphabet, no?

All electronics is designed to manipulate a flow of electrons. Electrons have mass and volume so you can almost think of electrons in circuits as water flowing through plumbing. The analogy is amazingly helpful if you think about it. The more electrons you have in one place, the higher the voltage. The more electrons moving together, the higher the current. The same as with water.

Power

Power is simply the energy required to do something. If you are moving a large amount of electrons, and moving them through something that is resistant of that movement, power is used. Power is voltage times current. Power is also voltage squared divided by resistance.

$$P = I * V$$

$$P = (V^2)/R$$

Ground and Source

Source is the positive part of your circuit. The plus end of your battery would go here. Ground is the negative node of your circuit. When you design your circuit, imagine a flow of electrons coming from the source, and heading to the ground. In reality electrons move from ground to source.

Now think of this as water. Water flows down the easiest quickest path between these two points. More resistance to flow less will flow.

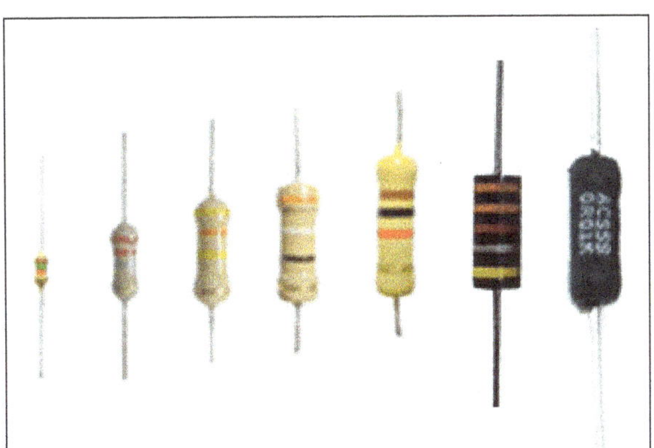

Resistors

These do exactly what they say. They resist the flow of electrons. These are necessary for several reasons:

- They can control how much current goes down each wire.
- They can control power usage.
- They can control voltages (since current, resistance, and voltage are interrelated).

The last point is important as it is the basis of Ohm's law, V=IR. Voltage = Current x Resistance. For example, suppose you take a resistor and connect the two ends of a battery with it. You know that your battery is 9V (or whatever) and you know the resistor is 3Kohm (determined by the color stripes on the resistor), so 9V divided by 3Kohm is .003amps (3 milliamps). So why is this information useful? Well now that you know the current, you can determine other useful things such as power. P=IV. You will notice that if you increase resistance, you decrease current. If you decrease current, you decrease power use. Put a 1ohm resistor between the battery and it will get so hot it could burn because of the power use. Use a 100Kohm resistor and almost no power at all will be used.

So about determining the value of a resistor, all resistors have the value labled on them. You will notice colored stripes on the resistor. Each stripe means a certain number.

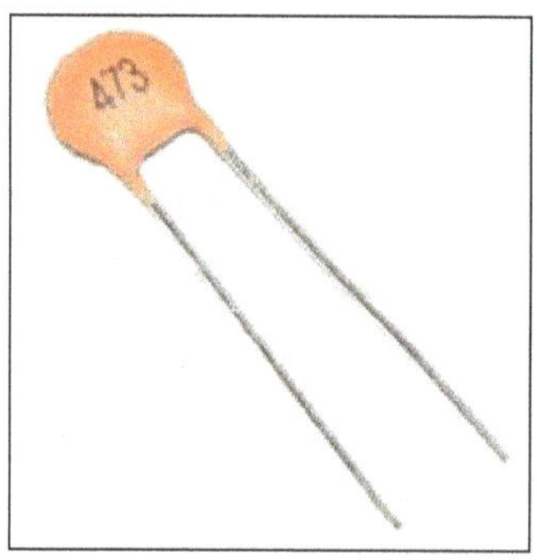

Capacitors

Now suppose you want to control how the current in your circuit changes (or not changes) over time. Now why would you? Well radio signals require very fast current changes. Robot motors because current fluctuations in your circuit which you need to control. What do you do when batteries cannot supply current as fast as you circuit drains them? How do you prevent sudden current spikes that could fry your robot circuitry? The solution to this is capacitors.

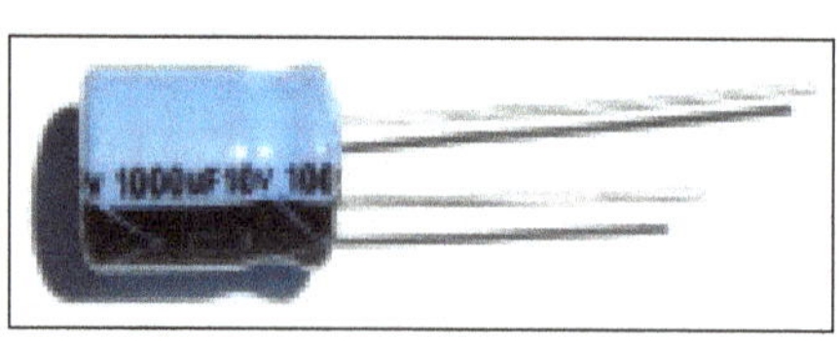

Capacitors are somewhat complex in theory, but most people can get by on the basics. Capacitors are like electron storage banks. If your circuit is running low, it will deliver electrons to your circuit. If your circuit is in excess (such as when your robot motors are turned off), it will store electrons. In our water analogy, think of this as a water tank with water always flowing in, but with drainage valves opening and closing. Since capacitors take time to charge, and time to discharge, they can also be used for timing circuits. Timing circuits can be used to generate signals such as PWM or be used to turn on/off motors in solar powered BEAM robots.

Quick note, some capacitors are polarized, meaning current can only flow one direction through them. If a capacitor has a lead that is longer than the other, assume the longer lead must always connect to positive.

How do capacitors charge over time? This Capacitor Charge Curve Chart should help. The discharge rate would be the direct inverse. Theoretically (as made obvious by the graph) a capacitor can never be fully charged or discharged, but in reality this is never the case.

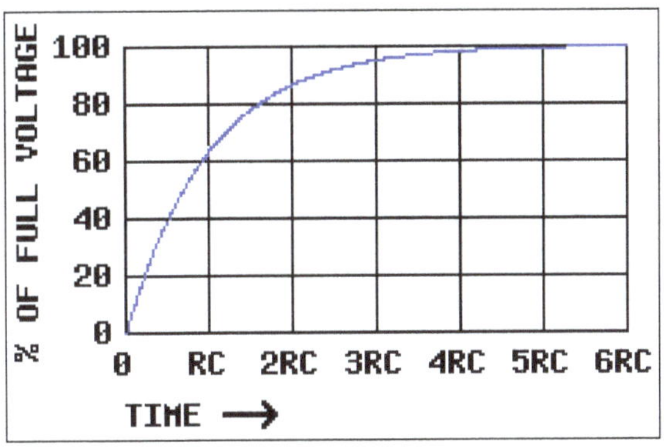

Power Surge/Drainage Management

The problem with using robot components that drain a large amount of power is sometimes your battery cannot handle the high drain rate. Motors and servos being perfect examples. This would cause a system wide voltage drop, often resetting your microcontroller, or at least causing it to not work properly. Just a side note, it is bad to use the same power source for both your control circuitry and your motors. So don't do it.

Or suppose your robot motors are not operating at its full potential because the battery cannot supply enough current, the capacitor will make up for it. The solution is to place a large electrolytic capacitor between the source and ground of your power source. Get a capacitor that is rated at least twice the voltage you expect to go through it. Have it rated at 1uF-10uF for every amp required. For example, if your 20V motors will use 3 amps, use a 3uF-30uF 50V rated capacitor. Exactly how much will depend on how often you expect your motor to change speed and direction, as well as momentum of what you are actuating. If your capacitor is too large, it makes take a long time to charge up when you first turn your robot on. If it is too small, it will drain of electrons and your circuit will be left with a deficit. It is also bad to allow a large capacitor to remain fully charged when you turn off your robot. Things could accidently short and fry, such as curious kitties that get too close. So use a simple power on LED in your motor circuit to drain the capacitor after your robot is turned off. If your capacitor is not rated properly for voltage, then can explode with smoke. Fortunately they do not overheat if given excessive amounts of current. So just make sure your capacitor is rated higher than your highest expected.

Capacitors can also be used to prevent power spikes that could potentially fry circuitry. Next to any on/off switch or anything that that could affect power suddenly should have a capacitor across it?

Capacitors can eliminate switch bouncing. When you flip a mechanical switch, the switch actually bounces several times within a microsecond range. Normally this is too small of a time for anyone to care (or even notice), a microcontroller can take hundreds of readings in a single microsecond. So if your robot was counting the number of times a switch is flipped, a single flip can count as dozens. So how do you stop this? Use a small ceramic capacitor! Just experiment until you find the power capacitance value.

Capacitors can improve efficiency and longevity of electric motors up to 100%. Place a small ceramic capacitor of like 10uF across the two leads of your motor. This works really well with el-cheap-o motors. Not much effect with high-end expensive motors however. These capacitors will also significantly reduce EMI (Electro Magnetic Interference) and system noise too.

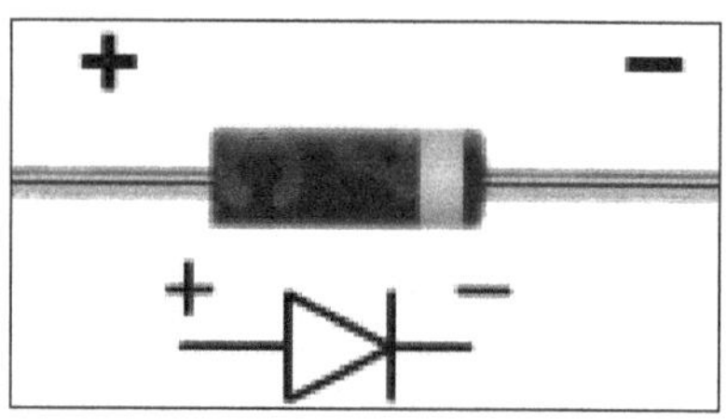

Diodes

Diodes are what you would use to ensure current flows in only a single direction. A great water analogy to a diode is a dam. Water never flows up a damn. But the analogy goes even further. With diodes, there is always a voltage across it (typically .7V forward voltage). Meaning if you have a diode come after a 7.2V battery, the voltage would then be 6.5V. This is just like a dam in that the water level will always drop. Doesn't current already always flow in only a single direction? No. RC circuits, or circuits involving AC power, or circuits that are noisy (such as with motors), involve currents that changes directions. So why you would only want current to flow in a single direction? Many many reasons. But for a beginner, you need to protect your circuitry from noise. A microcontroller would fry if current went the wrong way. Motor drivers and MOSFETs would too. Diodes are also useful for dropping high voltages to a lower more usable voltage.

Chart represents the current vs. voltage curve typical of diodes. The current passing through a diode changes non-linearly as voltage changes linearly.

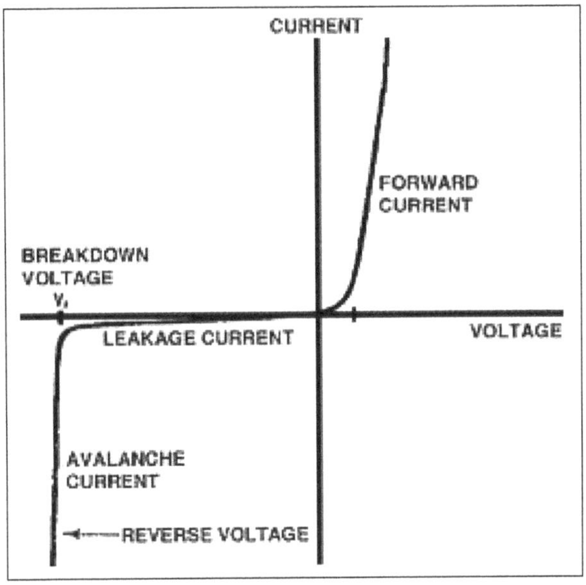

There is another special diode called a zener diode. With the water analogy, a zener diode is like a dam, but with a pump at the bottom pumping water back to the top. Zener diodes allow current to flow in reverse as well as forward. The forward voltage is still around .7V, but there is a different reverse voltage of around negative ~2.3V. You will probably never need to use a zener diode.

Mechanical Components

A robot consists of the following components: mechanical structure, transmissions, actuators, sensors, and controller. Although the components used in the robots are not exclusive to these (machine tools and many other machines use similar technologies), the high performances that are demanded of the robots require that they use components with specific functions.

The physical components of most industrial robots have some similarities with the structure of the human body. So, sometimes, people refer to the different components that make up the robot with terms such as waist, shoulder, arm, elbow, wrist, etc.

The basic components of a robot are as follows:

1. Locomotion System/Manipulator: The locomotion system determines the robot's movement. Whether it is rotatory motion, translator motion (full movement of the robot), etc. With this system, the robot can move right, left, forward, backward, climb up/

down, etc. This is accomplished with the help of a device that converts electrical energy into mechanical energy. These devices are known as actuators, and the most common one is the DC Motor.

Mechanically, the manipulator is the main component of the robot. It is formed by a series of solid structural components or joints that allow a relative movement between every two consecutive links. The parts that make up the manipulator are called the body, arm, wrist and actuator (or terminal element). The latter is commonly known as a gripper, claw, or clamp.

Robotic manipulators are divided into two categories; each has a unique function:

- Arms and Body: these parts of a robot are used for movement and positioning parts or tools within a work envelope. The arms and body are formed from three joints connected by large links.
- Wrist: the wrist consists of two or three compact joints, and it is used to orient the tools or the parts at the work location.

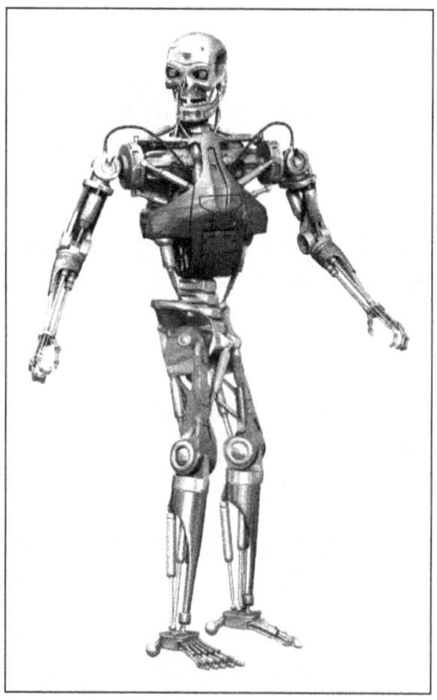

Robot manipulators consist of a sequence of link and joint combinations. The rigid components connecting the joint or axes are known as the links. The axes are the robot manipulator movable components that cause relative motion between adjoining links. These mechanical joints of the robotic arm manipulator are made up of five basic types. Three of the joints are rotatory types, and it is where the relative motion involves rotation between links. While two are linear, and they help in non-rotational motion between adjacent links.

Furthermore, the arm and body components of the robotic manipulators have four configurations, each of these components provides a different work envelop and is suitable for different application. They are gantry, cylindrical, polar, and jointed arm.

2. Controller: The controller is referred to as the physical brain of the robot, and it is a microcomputer with a Central Unit, memory, power and interfaces to access the external elements. This controller regulates each of the manipulator's movements, actions, calculations and processing of information. It receives and sends signals to other components (via input/output signals) and stores programs. There are several degrees of control, which are a function of the type of parameters that are regulated, and this gives rise to the following types of controllers:

- Position controller: It only interferes with the control of the position of the terminal element.
- Kinematic control: it regulates speed.
- Dynamic control: It controls the dynamic properties of the manipulator, motors, and associated elements.
- Adaptive control: helps the manipulator change and adapt to different positions.

3. Power Supply: A robot needs a power supply system to work. The power supply acts as a food to the robot, if you don't feed the robot; there is no way it can work. Therefore, the robot needs a power supply to work, and most robots use DC supply although solar power is becoming a focus for some types of robots.

4. Actuator System: Actuators are devices that help the robot with locomotion. In other words, they are components that convert the energy supplied to the robot into a physical result that can be used for movement.

There are a wide variety of actuators in robotic arms. There is the synchronous actuator and the asynchronous actuator. In the synchronous actuator, the motor consists of a rotor, which rotates synchronously with the oscillating field or current. In the asynchronous actuator, the motor is designed to slip so as to generate torque. Examples include Traction Motor, Hydraulic, AC Servo Motors, and Pneumatic.

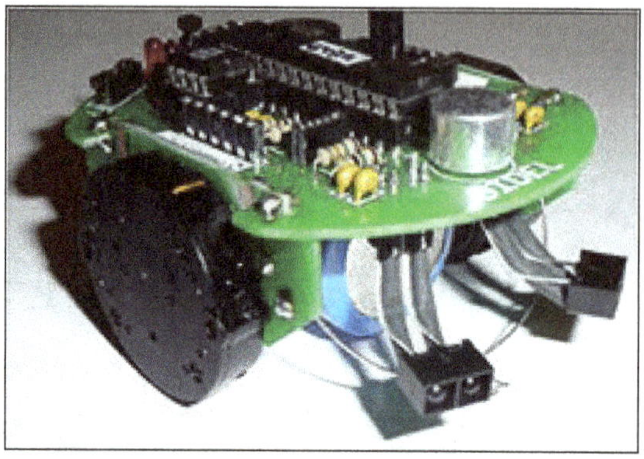

5. Sensor System: In a robot, the perception is ensured by the use of sensors. Sensors allow the robot to detect the conditions of the environment and can, according to its programming, respond to changes in conditions, obstacles, etc. Sensors provide information about the environment or internal components (e.g., motor or cylinder position, LED status). This information is used to calculate the appropriate command to be sent to the actuators.

There is a wide variety of sensors. Examples include Ultra-sonic, infrared or LASER rangefinders. These are the basis of laser scanners enabling the robot's central processing unit to become "aware" of its environment.

6. End Effector: An end effector in robotics is the device or tool at the end of the robotic arm. It aims to interact with the work environment. End effectors vary and depend on the application of the robot. The feet of a humanoid robot or wheels of a mobile robot are not regarded as end effectors; they are considered as part of the robot's mobility.

This is just a high-level overview of the common components of a robot. Of course as the field continues to develop the level of sophistication of these components, the types of components and their purposes will continue to evolve.

Materials

Robots inhabit the physical world, and various engineering materials are employed to provide shape, strength and durability, and to support non-structural components. Metals, plastics, and composites tend to dominate the structural elements, but other materials are occasionally used. The following segment summarizes some design-relevant properties. Most of the discussion concerns rigid materials, which constitute most structural elements in current robots. The final segments contain brief descriptions of elastomers and flexible tensile elements, mechanically distinct material classes with useful properties.

Dem bones.

Metals

Metals are strong, rigid, hard, tough, heat resistant, and isotropic (their properties have no directional dependence). This combination of characteristics is unmatched by any other common class of materials. They also tend to be heavy, and moderately to extremely difficult to fabricate. Metals are typically shaped by forging, bending, machining (sawing, drilling, milling turning), and grinding. Some can also be cast, though typically some additional machining is required. Metals in their pure form are often surprisingly soft, so engineering materials are typically alloys with other elements. The amount required is often remarkably small: less than 1% carbon for steel, a few percent copper or magnesium for aluminum can produce nearly an order of magnitude increase in yield strength over the pure material.

Metals are basically unmatched in their combination of strength (hardness) and toughness. Other very hard and strong materials tend to be brittle, which makes employing their underlying strength a challenge. What gives metals this property is the fact that they deform before they break. The stress at which a metal deforms is known as its yield point, and represents a practical upper limit to the forces that can be applied to a metal structural element. However, this deformation greatly reduces the stress concentration at the tip of an incipient fracture, and prevents brittle propagation of the defect. Some plastics also have this property, but with much lower yield points.

The following metals are some of those most commonly encountered as structural materials:

Steel

Steel is the most common and least expensive metal, and also one of the strongest. Unhardened mild steel yields at 30,000-50,000 psi. For structural purposes, it can be hardened easily to 100,000 psi, and for tooling, can be strengthened to nearly 300,000 psi. It has a density of approximately 8 times the density of water (7.9 gm/cc), and a melting point around 1400 degrees C (pure iron 1530 °C), which looks white hot.

Making steel.

Ordinary, or low alloy steel is mostly iron, with a small amount of carbon (less than 1%) serving to harden it. It often also contains 1% or so manganese, and small amounts of other elements such as silicon. A 4-digit AISI/SAE code such as 1015 or 1040, or an analogous 5-digit UNS number (10150, 10400) is often used to describe steel alloys. The first two AISI digits specify a broad type; the initial 10 in 1015 means ordinary low-allow carbon steel. The last two digits are the carbon in hundredths of a percent.

In general, the more carbon a steel has, the more it can be hardened through heat treatment, which is the process of bringing it to orange heat, and then cooling it rapidly (usually in water or oil) sometimes followed by a milder re-heating or "tempering" process, which makes the hardened material less brittle. "Annealing" is a process where the steel is brought to red or orange heat, and then cooled very slowly (over a period of hours or even days), which returns it to a (comparatively) soft state. Above one percent carbon, it becomes increasingly difficult to control brittleness, and steels grade into the material known as cast iron.

Steel stock comes in hot-rolled and cold-rolled forms. Hot roll is typically the cheapest and has heavy, black "mill scale" oxidation on the surface. It may also have variable surface hardness due to rapid cooling from contact with the roll. These properties can be tough on tooling, causing more rapid wear or breakage than expected. Cold roll is worked in a cooled state, and has a more finished surface and more uniform consistency. It tends to be a bit harder and stronger than hot roll due to work-hardening effects, but generally can still be machined.

Mild or annealed steel is the only form which can be worked. It can be machined with moderate effort. Machinists tables say that it can be cut with high-speed-steel (HSS) tools at up to 100 surface feet per minute (that's the rate at which the tool moves across the metal surface) but this is a bit of a maximum with adequate cooling, lubrication, and appropriate feed rates. As beginners, if we want to avoid damaging our tools, 25 SFM is a good point to aim for. It looks slow, but it will get the job done.

Steel is easily joined by welding (if you have a welder), and the resulting joints can be as strong as the underlying material. It can also be joined by a lower temperature process known as brazing, but you still generally need a welding torch. Tinned sheet steel (the material of the old tinsmiths) can be joined by low-temperature soldering, but it takes some practice to master and much higher-power equipment than is used for soldering electrical components. For our robots steel will be useful for fasteners, bearings, springs, and other small parts, but is probably overkill for most of the structural components.

Aluminum

Aluminum is roughly one third as heavy, and one third as strong as steel (density 2.7 gm/cc, yield point 10,000-40,000 psi). It generally melts around 600 °C (pure aluminum 660 °C), which is just above red heat. It is generally malleable and ductile, and takes a high polish. Engineering aluminum is usually an alloy containing a few percent other metals which provide strength and hardness. Common additives are copper, magnesium, silicon, and manganese.

Aluminum does not rust, but sometimes corrodes in wet environments, especially if there are acidic or basic impurities in the water. Surface treatments generically referred to as "anodization" can provide a high degree of protection. Chemically, aluminum is unstable both with respect to air and to water, but self-protects with a very tough, microscopically thin layer of aluminum oxide. In pure oxygen, it will burn intensely, but fortunately it is very difficult to ignite in air, even in fine shavings.

Aluminum is MUCH easier to machine than steel, which makes it the structural metal of choice for our robots. It cuts easily with HSS tools at 100 SFM. Aluminum has a tendency to "gall", producing a rough surface if the tooling is not sharp. It also sometimes sticks or "melts" to the cutting edge of the tool. These accumulations can usually be snapped off a cooled tool using a piece of hard plastic, soft metal or even a fingernail, but you need to be careful not to chip the cutting edge or cut yourself in the process. Aluminum cannot be joined by soldering, and can be welded only with special techniques. Components are hence generally assembled using threaded fasteners.

For small components, aluminum is about twice the cost of steel. For large components, the difference is greater. However, the easier machining often more than makes up for the difference in the cost of the bulk material.

Aluminum is non-magnetic, and an excellent electrical and thermal conductor. Its use as an electrical conductor is usually limited to applications where very large conductors are needed, due to the difficulty of making good connections (it cannot be soldered and contact surfaces sometimes become resistive and heat up). It is the material of choice in heat sinks and heat exchangers except in applications where corrosion resistance is needed or temperatures above 300 °C will be encountered.

Copper, Brass and Bronze

Brass and bronze are copper alloys occasionally used for structural purposes. Brass refers to an alloy of copper and zinc, bronze to copper and other elements, most commonly tin, but sometimes, aluminum, silicon, or other metals. They have about the same density as steel, and generally half to two-thirds its strength (density 8.4 gm/cc, yield point 25,000-60,000 psi). Both melt at around 900 °C, a good red-orange heat. Pure copper melts at 1080 °C, bright orange.

Copper alloys are non-magnetic, non-sparking, and some of them are very corrosion resistant. They are used instead of steel when these properties are important enough to justify the cost, which is 4-5 times that of normal carbon steel in small amounts and more in large pieces. Brass is sometimes used in bearings as it has self-lubricating properties both against itself, and against steel. An unlubricated steel-on-steel bearing will rapidly self-destruct, as will aluminum (which is generally not hard enough for bearing purposes in any case).

Most common brasses machine easily. The machinability of bronzes varies from easy to difficult. Many copper alloys can be joined by soldering.

Pure copper is rarely used structurally as it is very soft, but it is used nearly universally as an electrical conductor, and sometimes in heat exchangers in applications where aluminum would corrode. It is used in plumbing due to the combination of corrosion resistance, strength and heat resistance (compared to plastics), and easy solderability.

Stainless Steel

Technically stainless is a variety of steel since iron is the largest component. However, it contains relatively large amounts (10%-20%) of both nickel and chromium, and sometimes other elements such as molybdenum or vanadium. These provide a high degree of corrosion resistance. Its strength characteristics, density, and melting point are similar to carbon steel. Stainless can be magnetic or nonmagnetic. Non-magnetic stock may become magnetic when bent or deformed, which should be remembered if you are counting on non-magnetic properties. It is a relatively poor conductor of heat and electricity compared to copper and aluminum, and even to regular carbon steel. (But it is still a really good conductor compared to the human body or insulating materials, so take the same care around electricity you would with any other metal).

Stainless is used as a structural material mainly when corrosion resistance needs to be combined with high strength and temperature resistance. Machined stainless has an attractive look and will take a mirror polish. It is scratch and chemically resistant and easy to clean. Because of these properties it is sometimes used where appearance is important. Stainless typically costs 2-3 times as much as carbon steel for small components, more for large ones.

Stainless steel is notoriously difficult to machine. Even when annealed it work-hardens rapidly when cut. It has a tendency to break or chip tooling. If you must machine it, use carbide tooling if available, and run at 1/4-1/3 the speed you would cut carbon steel of equivalent hardness. Stainless cannot be soldered or brazed, and needs special techniques for welding. Correctly made, welded joints are of high quality.

Titanium

Titanium is an exotic metal sometimes found in high-tech applications. It is about half the density of steel (4.3 gm/cc) but some alloys can be made almost as strong (yield 15,000 + psi). It is high-melting (1670 °C), extremely corrosion resistant, and biologically inert. It is also very expensive, difficult to machine (more so than stainless), and can only be cast, forged, or welded with very specialized equipment. Thin shavings will burn in air if ignited, so turnings are a fire hazard. It will also burn in carbon dioxide and even pure nitrogen, which makes extinguishing such a fire tricky (dry sand will work).

Titanium is generally used only when biological inertness is essential, or the application requires an unusual combination of light weight and high strength (e.g. some aerospace applications).

Other Metals

1. Magnesium has properties similar to aluminum, but is lighter (density 1.7 gm/cc), more expensive, more corrosion prone, and its shavings will burn intensely in air (machining it is a fire hazard). It is used only when the low weight outweighs all the negatives. We probably won't use it.

2. Zinc is sometimes used for low-cost metal components in cars, locks, toys, and low-quality machinery. The advantage is that components can be quickly die-cast at low temperature. The down side is that the material is weak and not very heat resistant for a metal, and nearly as heavy as steel (density 7.1 gm/cc).

3. Lead is also easy to cast at low temperature, and some of its alloys have better mechanical properties than zinc. It is also traditionally used to make weights due to its high density (11.3 gm/cc). Some of the best low-temperature solders contain lead. Unfortunately some of its compounds are toxic if eaten or inhaled, and can cause cumulative brain damage in children at low dosages. The extensive use of lead carbonate as a white pigment in house paint in the 19th and early 20th centuries resulted in widespread exposure as deteriorated paint chipped and peeled directly into living spaces. Lead thus became widely known as a very bad thing, and most uses have been discontinued. Some

electrical solder still contains lead as it is easier to use than any of the lead-free alternatives so far discovered. Large amounts are still used in automotive batteries. Use lead solder only in well ventilated areas. Don't eat solder or batteries, and wash your hands after handling them.

4. Mercury (not a structural metal) is primarily interesting because it is liquid at ordinary temperatures. It was widely used in household thermometers and in mercury switches, which provide a very nice spark-free, position-sensitive electrical contact. Mercury is significantly more toxic than lead, and the free element is a hazard because it evaporates slowly into the air and can be inhaled. If you drop a mercury switch, the glass tube can shatter and then little droplets of mercury will race all over the floor and into every crack and corner. UR would have to send in a decontamination team, and we would probably be closed down forever. So I'm not putting out the old mercury switches we got once as robot orientation sensors. Interesting factoid: mercury is denser than lead (13.5 gm/cc).

5. Depleted uranium is even better for weights than lead or mercury (density 19.1 gm/cc). It is used to ballast the keels of some high-performance sailboats and in military armour-piercing bullets. Unfortunately it is slightly radioactive and can cause lung cancer if inhaled as dust, and probably other cancers if eaten in a bio-available form. Fine shavings will burn in air if ignited, producing uranium oxide as finely divided, and inhalable smoke. We would need to get special permission to use it in a robot.

Plastics

Plastics are mainly 20th century products of applied chemistry, and they provide materials with an amazing array of properties. No existing plastic approaches metal in absolute strength and heat resistance, but otherwise they can be produced with practically any mechanical characteristics seen in other materials.

Chemically, plastics are polymers: long chains of repeating subunits. Most often these units are organic compounds, but there are some plastics based on silicon and other unusual chemistry. The chemicals representing constituent subunits are known as monomers. The properties of a polymer however, are most closely associated with attributes of the chain, especially its length, its local flexibility, and whether chains are bonded to each other - an attribute known as crosslinking.

Plastics as a group tend to be waterproof, corrosion proof, and resistant to chemical attack particularly by acids and bases. Some can be softened or dissolved by organic solvents. Many plastics can be made in transparent form, and these form a useful class of optical materials. In general they are excellent electrical insulators and relatively poor conductors of heat.

Most plastics are subject to photo-degradation by ultraviolet radiation, which has sufficient energy to disrupt carbon-carbon bonds and thus break the polymeric chains.

Exposed to outdoor light they cloud, yellow, weaken, and eventually disintegrate over a period of weeks to years. Plastics intended for outdoor use have ultraviolet inhibitors added to slow the process. Some of these are good for a decade or more in the sun. Long term however, the only way to prevent eventual degradation is to block exposure to UV light.

Most plastics soften at relatively low temperatures; many lose all or most of their structural strength above 100 °C. A plastic that retains useful mechanical properties above 300 °C is a rarity.

Plastics are widely used structurally as rigid bulk material. The best engineering plastics are approximately an order of magnitude lighter, and an order of magnitude weaker than mild steel (yield strength 3,000-12,000 psi). In weight-limited applications, this makes them an attractive alternative. They also find application as flexible films and fibers. In this form their strength to weight ratio can greatly exceed that of steel. Polymers are also the basis for most elastomers and a large class of semi-rigid materials.

Following are some of the plastics most frequently used as rigid structural materials:

Polystyrene

Polystyrene is mostly included here as a plastic NOT to use in robotics. It is cheap, rigid, and easily injection molded. It is widely used in cheap toys and plastic model kits, and as expanded "styrofoam". Chemically, it is relatively inert, but can be dissolved in acetone and some other organic solvents. This permits easy and permanent bonding with plastic model cement. However, styrene is relatively weak as rigid plastics go (yield point about 3,500 psi), and particularly prone to brittle fracture. It can be machined with modest care. Drilling must be done slowly to prevent melting or fracture. It has a tendency to fracture when tapped or if screws are over tightened. The main application of polystyrene in robotics is as a remarkably inexpensive source of detailed and complex pre-made parts scavenged from toys or model kits. Styrofoam materials are sometimes useful structurally when extremely light components are needed. If constructed with a tensile skin, such parts can be remarkably strong for their weight.

Plexiglass

Also known as acrylic, and polymethylmethacrylate, plexiglass is widely available as sheets intended as a non-shattering replacement for glass windows. Although certainly less prone to impact fracture than sheet glass, plexiglass is not a particularly robust engineering plastic. Its absolute tensile strength is quite high (yield point around 10,000 psi), but like styrene, it is very brittle. A hard blow will cause fracture, as will strong localized pressure e.g., from a threaded fastener. Also like styrene, it has a tendency to fracture or melt if drilled too fast. It is a satisfactory material for small to moderate sized transparent panels that are not expected to sustain strong impact. Plexiglass is a relatively inexpensive plastic, less than the cost of aluminum in bar form, and considerably less in large commodity sheets designed for windows.

Rigid PVC

PVC is short for polyvinylchloride. It is most familiar as the soft, flexible plastic that covers wires, and from which cheap raincoats, inflatable toys, and imitation leather is made. Without plasticizers however, it is a strong, rigid material. Plastic plumbing components are often made from PVC. Its yield strength is around 7,000 psi, and it is considerably more impact resistant than plexiglass. It also deforms 10-20% before breaking. These properties make it a reasonable structural material. It machines well, and can be permanently glued together using solvent welding cements. PVC is the least expensive of the structurally useful plastics, less than half the price of aluminum stock in small pieces, and very much cheaper in large amounts of commodity material. It can be obtained in rod or sheet form in large sizes. PVC pipes are readily available at low cost in the plumbing section of hardware stores, and can often be used as structural components.

ABS Plastic

Short for Acrylonitrile-Butadiene-Styrene, ABS is a copolymer formed from three different monomers. It is an example of an engineered plastic where monomers whose homo-polymers have different characteristics are combined to produce a material with new, desirable properties. In this case a synthetic rubber, polybutadiene becomes interlaced with a rigid styrene/acrylonitrile copolymer. The product has the rigidity of polystyrene but is not brittle, and polar attraction from triple-bonded nitrogen in acrylonitrile provides increased inter- (and intra-) molecular bonding. The resulting plastic is strong (yield around 5,000 psi), and very impact resistant. ABS exhibits considerable ductile deformation, stretching 20% or more before breaking. It is what LEGO bricks are made of.

ABS machines smoothly and easily. It can be tapped for threaded fasteners, and holds them well. It can be effectively glued or solvent welded. It retains its strength and impact resistance at temperatures down to -40 °C. On the negative side, it loses its strength above 80 C, so is not good for hot locations. It is also not readily available in transparent form. In small amounts, ABS costs about the same as aluminum. In large commodity forms (e.g. some plumbing pipe) it is considerably cheaper.

Polycarbonate

Also known by the trade name “Lexan”, polycarbonate is the classic “indestructible” miracle plastic. It is very strong (yield point around 10,000 psi), extremely impact resistant, and deforms up to 100% before breaking. Sheet material is sometimes formed by cold bending, just like a metal. It can be made in crystal-clear form, and is popular for “bulletproof” windows. It is resistant to acids, bases, alcohols, and most oils, but can be softened by acetone and some other solvents.

Polycarbonate machines reasonably well. It has some tendency to melt when drilled, so drilling should be done at low speed, and the tool withdrawn frequently so that chips do not become trapped between the drill and the whole wall where they turn instantly into a gooey mess. It can be glued, though not as robustly as PVC. Polycarbonate is one of the more expensive plastics, 1.5+ times the cost of aluminum in small amounts. Unlike PVC, plexiglass, and ABS it is not commonly used in household plumbing or other high volume consumer applications, so cheaper commodity forms are not generally available.

Acetal

Also known by the trade name Delrin, acetal is effectively polyformaldehyde - a simple backbone of alternating carbon and oxygen atoms. It is strong (yield 10,000 psi), relatively ductile (up to 60% deformation) and very chemically inert. It is unaffected by acids, bases, and all common solvents. It is abrasion resistant and has a low coefficient of friction, which makes it useful for low-pressure bearings and sliding parts. It is somewhat susceptible to impact fracture, but resists brittle fracture under gradually applied forces.

Acetal machines easily, can be tapped, and holds screws reasonably well. In applications where screws might tend to loosen, the low coefficient of friction could be a complicating issue. Due to its slippery nature, it is less problematic to drill than stickier plastics, but it is still possible to melt or burn a hole by too high a drill speed. Because of its unreactive nature, acetal cannot be securely glued, though some flexible adhesives may provide a low-strength bond. Acetal is relatively inexpensive - a bit less than aluminum in small amounts. It is not a household commodity.

Polyethylene and Polypropylene

Polyethylene (PE) and polypropylene (PP) are common, inexpensive plastics with similar properties. Both are relatively soft, with considerable ductility and ultimate bulk strength around 5000 psi. They are slippery, non-gluable, chemically inert, and unaffected by all common solvents. Because of these properties they are commonly used as containers for all types of chemical reagants from strong acids to paint thinner. Small non-polar organic molecules will slowly diffuse through both plastics, so they are not suitable for extremely toxic liquids. Huge amounts of both plastics are used as disposable food and beverage containers. Large slabs of both PP and PE can sometimes be found as consumer products in the form of cutting boards.

Both plastics have enough mechanical strength to be useful structurally, though they are considerably softer and more flexible than the other engineering plastics we have considered. They can be machined with reasonable care. Some precautions are needed because the softness and flexibility of the material. Tools must be very sharp, and material must be well supported so that it does not bend away from the tool. Neither material can be glued, and even flexible adhesives often hold only weakly. PE and PP are both inexpensive, only slightly pricier than PVC.

Polyethylene actually comes in three forms. Low density PE (LDPE) is rather soft and susceptible to both impact fracture and fatigue when employed in bulk. This limits its usefulness as a structural material. High density PE (HDPE) is much stiffer and more impact resistant. It is often used for large plastic containers such as garbage cans and chemical drums, and is quite durable. Ultra High Molecular Weight PE (UHMWPE) is a relatively recent development. It is one of the most impact and abrasion resistant plastics known. It has been used for iceless skating rinks and joint replacement implants.

UHMWPE can be drawn into a fiber known by the trade names Spectra and Dyneema. The ultra-long molecules, when brought into alignment give these fiber almost incredible properties. Its absolute tensile strength can exceed 350,000 psi, which is higher than very high strength steel cable, and has been exceeded by metals only in laboratory settings. In terms of strength to weight ratio, it is 10 times as strong as steel cable. The only down side is that the fiber loses its strength above 100 C. It is used in bullet-proof clothing and other military applications. Amazingly, you can buy this material at relatively low cost in your local sporting-goods store - as fishing line.

Nylon

Nylon is the original DuPont trade name for its revolutionary fiber. The term is now used generically for synthetic polyamide plastics. Although the word is usually associated with fibers and fabric, nylon is also used in bulk solid form to make structural components. It can be very strong, with a tensile strength as high as 12,000 psi, and it undergoes considerable plastic deformation before breaking. It does not have the impact resistance of ABS or polycarbonate, and when repeatedly stressed to near the elastic limit, eventually fatigues and breaks. Kept well under the elastic limit, it can be repeatedly stressed almost indefinitely. It has a low coefficient of friction (i.e. it's slippery), resists abrasion, and can be easily injection molded into highly precise forms. This combination of properties makes it an ideal material for plastic gears and other moving parts. It is far less expensive to produce nylon gears than metal ones. Most of the gear trains and sliding components in modern consumer devices are made of molded nylon.

Nylon can be machined drilled at tapped with the usual precautions against melting needed with plastics. It cannot be glued. It is relatively inexpensive, slightly less than the cost of aluminum in small amounts.

PETE

PETE is an abbreviation for PolyEthyleneTErephthalate, a polyester plastic. It is best known as the super-strong stuff from which soda bottles are made, but it is also the basis for Dacron polyester fiber, and Mylar film. It is sometimes used in bulk form, but is of most interest to robotics as a very strong sheet and film material. The strength is produced by stretching the amorphous plastic in both “x” and “y” directions. This stretches out the polymer molecules so they are aligned with the sheet surface. Any further deformation has to break the attraction of long molecules along their entire length, rather than just locally as would be the case if they were randomly coiled up like a pot of spaghetti. The result is great strength and stiffness in the two extended directions of the film. PETE can be produced in a form that withstands temperatures above 200 °C. This material has been used to make plastic food containers that can be placed directly in an oven for heating.

Because of its strength, durability, and chemical inertness, PETE film has been used in a huge number of applications from packaging, to magnetic tape substrates, to balloons. The familiar shiny surface of Mylar is produced by vacuum deposition of aluminum vapor. With appropriate adhesives, the film can be securely glued for assembly.

Composites

Technically, a composite is a structural material composed of two or more simpler substances, often having very different properties, combined (in macroscopic manner) to yield a material with combinations of properties not present in any of the components in isolation. For example, very strong, but brittle glass is combined with relatively weak, but more flexible plastic resins to yield fiberglass which is both exceptionally strong and extremely fracture resistant.

Most engineering metals are technically composites. Carefully polished and etched, and under modest magnification, alloys can be seen to consist of different crystal phases interlocked and cemented together. It is largely the combination of the different mechanical properties of these components that gives alloys their desirable characteristics. We will use the term in its more generic sense for materials made by combining more obviously disparate elements.

A characteristic of many composites is non-isotropy in mechanical properties such as strength. A carbon-composite fishing rod can endure lunges from a hooked marlin that would bend steel in the same form, but split if stepped on. This is due to a common composition technique where a strong but brittle material is made into thin fibers which are bonded together along their length with weaker, more flexible “glue”. Brittle fracture propagates because of the mechanical concentration of stress at the tip of a fracture supported by the stiff bulk of material behind it. Individual fibers can easily tolerate the (minor) distortion of a stressed bulk element. Any concentrated forces because local yielding of the softer material, spreading the stress along the length of the fiber and avoiding the bulk mechanical leverage that propages brittle fracture. Along the fibers, the composite has most of the strength of the fiber material, since the weak glue holds the fibers together edge-to-edge over a long distance. Pulled apart across the fibers however, the composite has only the strength of the glue.

The engineering of modern composites is an entire field in itself. In building our robots, we are likely to encounter only a few, including wood (yes, wood), fiberglass, possibly pre-made carbon fiber elements, and glass-filled plastics.

Wood

Wood might not seem like an appropriate material for shiny modern robots, but it has its advantages. Chief among them are that wood is inexpensive, widely available, and easily formed with commonly available tools. But wood itself is a remarkable structural material. It is a natural composite (cellulose fibers glued together with a substance called lignin) having a stiffness to weight ratio that matches steel or aluminum. This is important in structures that are sufficiently extended that they deflect under their own weight. Wood is stronger in absolute terms than many plastics, (generally several thousand psi along the grain) and tougher than most. More importantly, once dried, it has little tendency to "creep", a slow deformation under stress that plagues many plastics. Wood can be securely joined together in more ways than practically any other material. It can be nailed, screwed, glued, bolted, wedged, doweled and dovetailed. There are dry assembly methods that are as secure as anything using external fasteners. And it can be finished to have practically any appearance desired.

Wood on the hoof.

Wood does have its problems. It is strong only in one direction - along the grain. It is typically only 1/100 as strong in tension across the grain, and is usually significantly weaker in compression as well. For structural purposes, wood should be considered to have effectively zero tensile strength across the grain. Plywood, which is made by gluing thin sheets together with the grain at different angles, has good two-dimensional strength. A strong isotropic wood-based material is not currently available. Particle board is somewhat isotropic, but it is weak in all directions. Theoretically fibers could interlace in all three directions, but trees have not managed it on a bulk scale. Certain "unsplittable" woods such as elm are known, but they are really not even close to isotropic.

Wood has knots, splits, defects, uneven grain, and variable strength and density - even within the same board. And it moves when the humidity changes. A tight fit on a dry day may buckle on a humid one. More commonly gaps open in originally well-fitting assemblies. Originally flat boards bend or twist. The movement is uneven. Along the grain wood changes little in size. Across the grain shrinkage of 5 percent or even more may occur, especially if it was slightly green to begin with. Much of the art of woodworking is involved with ensuring that articles remain functional in the presence of wood's natural movement.

For our robots the most useful forms of wood are likely to be small bars and dowels, which make reasonable lightweight beams, and some of the thinner plywood, which are useful as mounting substrates and simple enclosures.

Fiberglass and Carbon Fiber

Fiberglass is made by impregnating a matrix of oriented glass filaments with liquid plastic resin, usually an epoxy, which then sets, binding the whole into a strong and rigid material. The glass filaments may be longitudinal which yields a material with primarily one-dimensional strength like wood, or made into cloth or a felted mat, which gives two-dimensional strength. Three-dimensional weaves are possible, but rarely encountered, and usually only in expensive, exotic applications. Cured fiberglass composite can be as strong as medium steel in its strong directions, and is considerably lighter (density around 2 gm/cc).

Fiberglass is a reasonable do-it-yourself material. Glass fabric and epoxy resins are readily available at automobile parts stores, craft and hobby dealers, and even some hardware centers. It is one of the easiest ways to make curved custom parts. The hazards associated with the materials are manageable: the resins can be irritating, and glass fibers can sometimes penetrate the skin causing a rash and itching. Both need to be absolutely kept out of the eyes, and dust from sanding or sawing should not be inhaled or left where it could be disturbed and become airborne. If cured fiberglass must be abraded, wet-sanding methods should be used.

Carbon composite, sometimes called "graphite" is similar to fiberglass except that ultra-strong carbon fibers made by pyrolizing certain oriented polymers in a furnace are used instead of glass. It is both stronger and more expensive. Carbon fiber is more hazardous to work with than glass. The stiff fibers penetrate skin easily, and unlike glass, which can (very slowly) dissolve in the body, they NEVER go away. Inhaled carbon fiber is a possible carcinogen though most industrial fibers are in the 6-10 micron diameter range, which is significantly larger than the .5 micron diameter which is most damaging to the lungs. In any case, precautions should be taken. Premade carbon composite parts should never be sawed or sanded without respiratory protection and containment systems that prevent residues from escaping into the environment.

Filled Plastics

Plastic resins are sometimes mixed with (relatively short) fibers of harder or stronger materials to increase their yield strength. The most frequently used filler is chopped glass fiber, but other ceramics, metals, and even other plastic fibers are sometimes used. Because the fibers are short and generally do not have coordinated orientation, such materials are not as strong as engineered composites using the same materials (e.g. fiberglass). However, glass-filled plastics can easily have twice the strength of the un-modified resin. This is usually at the expense of ductility and sometimes impact resistance, although for brittle resins, glass fill can increase impact resistance by "bridging the gap" and impeding fracture propagation.

Ceramics

Ceramics can be thought of as artificial igneous rocks - that is, heat-fused refractory, fully oxidized materials. They are typically made by heating specially compounded mixtures of minerals to high temperature in a furnace, whereupon they stick together and remain firmly stuck after the material cools. The mixtures may or may not fully melt, and the process may also involve high pressures or special atmospheres. The canonical example is pottery, which has been known since antiquity.

Ceramics were not traditionally thought of as structural engineering materials. Except for bricks, which were not usually thought of as ceramics? This started to change in the mid-20th century when architects began using glass as a load-bearing material. Today high-tech ceramics are important structural materials in applications ranging from electrical insulators, to bearings, to gas turbine blades. The brittleness associated with glass and porcelain has been greatly reduced in many of these materials. Although ceramics do not yet match good metals in toughness, they can greatly exceed metals in strength, hardness, and wear resistance, especially at elevated temperatures.

Because of their hardness, ceramics are difficult to work. They are often directly manufactured in their final form, or are subject only to a final polishing step. We will probably not have much reason to use ceramics as structural materials in our robots outside of the occasional glass window.

Glass

Glass is most easily available in ordinary "window glass" form, which has familiar fragile properties including the tendency to shatter into large, extremely sharp pieces. Glass is intrinsically a very strong material. Its fragility is due to brittle fracture where a crack, once started, mechanically concentrates stress at its tip. Since glass has effectively zero ductility, the stress cannot be relieved by plastic deformation, and the fracture propagates rapidly. Glass surfaces generally contain microscopic scratches, and when that surface is placed under tension these scratches serve as initiation sites for fractures.

The fragility of glass can be reduced by various “tempering” procedures. Essentially, tempering places the outermost layer under compression, balanced by tension on the inside. This means that the glass can be stressed without placing the surface under tension, so microscopic scratches have no tendency to grow into fractures. The flip side is that if a chip or scratch penetrates the compressed surface layer, the pre-existing tension in the interior causes the entire volume to rapidly fracture into small pieces, as is observed in automobile side windows. In general such small pieces are less of a hazard to people than large sharp shards. Tempering was originally accomplished by rapidly cooling of the surface of glass in a hot, plastic state. The surface layer freezes, with shrinkage being locally accommodated by the still-plastic interior. Then as the interior cools, its shrinkage pulls the surface layer into compression. This technique really only works well for simple shapes such as sheets and smooth curves.

More recently, methods of “chemical” tempering have been introduced that can be used to treat arbitrary shapes. Here, glass in its solid state is placed in a chemical bath that replaces certain mobile elements in the glass matrix (e.g. sodium in soda-lime glass) with larger analogous atoms (e.g. potassium or cesium). Since these take up a bit more space, the surface layer is placed under compression. Usually this is done at elevated temperatures to increase the mobility of the atoms, but not high enough to soften the glass. One of the simplest methods is to place pieces of soda-lime glass to be tempered in a bath of molten potassium nitrate (about 350 °C = 660 °F) for 8-24 hours. You could almost do this in your home oven, but don’t. Molten KNO3 can react violently with organics and metals, and it decomposes with evolution of oxygen and nitrogen only slightly above its melting point (at 400 °C). Your home oven has hot spots that will cause a disaster.

Extreme tempering has recently been used to produce products such as “gorilla glass” (used in smart phone screens and other applications), thin sheets of which can be bent double without shattering, and basically don’t seem like glass at all.

Untempered glass sheet (e.g. window glass) can be fairly easily “cut” by scoring and breaking using an inexpensive glass cutter. Larger pieces can shaped into more complex forms, such as highly precise optical components by abrasive grinding and polishing, but this needs special wet-process equipment. Minor grinding can be done on an ordinary abrasive wheel or a stationary belt sander if the abrasive can be kept wet without creating an electrical hazard or damaging the equipment. Dry machine grinding of glass typically causes fracture from thermal stress. Sharp edges can be relieved using a hand sanding block and fine (>= 220 grit) sandpaper. Special carbide and diamond drill bits are available that, with some care, can be used to put round holes in glass sheet. The diamond hole saws can also produce a useful disk. They are always used with water or other liquid coolant. Tempered glass typically cannot be cut, and can be only lightly ground and polished. It must be shaped before tempering.

Window glass can be bought inexpensively at home centers and hardware stores. Specialty commercial and auto glass stores can provide plate glass in different thicknesses and colors. They can also order tempered glass sheets made to size and with holes and polished edges at extra cost. Optical houses such as Edmund Optics can supply high-quality glass lenses, ports and filters at very high prices (and at even higher price, some components made from non-glass materials such as sapphire and silicon).

Elastomers

Elastomers are rubbery materials. They have unique mechanical characteristics, the most useful of which is their ability to undergo extreme deformation under modest stress and then rapidly recover their original form when the stress is removed. In a bulk sense elastomers, like liquids, have a relatively high compressive modulus. Under uniform (isostatic) pressure, elastomers change very little in volume. Their interesting properties can be attributed to comparatively low shear moduli that effectively allow them to flow under stress. Elastomers are sometimes modeled as liquids with extremely high surface tension. This is a reasonably good model for a spherical rubber ball. Certain modifications need to be made to accommodate other shapes.

Elastomers are typically polymers, often cross-linked assemblages of loosely coiled chains. They are generally waterproof and can be made highly resistant to chemical attack. Their combination of elastic and conforming properties makes them ideal materials for seals and gaskets. Rubber O-rings can reliably seal a moving shaft against pressures of several thousand psi. They make hydraulic pistons possible. Elastomers are also used in shock and vibration reducing mounts, flexible connectors, and for surfaces requiring good grip, high friction, and a soft touch. Some elastomeric components can be deformed millions of times and still retain their original shape and resiliency (think tires).

Cured elastomers are extremely difficult to shape. Because they deform under low stress, they cannot be cut precisely. They tend to be abrasion resistant, and abrasive machining, if possible, often produces an unacceptably rough surface. Thus most elastomeric parts are manufactured in their final form. Some elastomers have an uncured form that can be melted and molded (e.g. unvulcanized rubber); others start as liquid resins that are polymerized directly in the mold.

In robots, elastomers are used to provide traction against the ground or friction on grippers. They are used as soft contacts to reduce impact forces and prevent damage to and by parts that contact the environment. They provide passive compliance in joints and manipulators, which is often essential for robust (and damage-free) interaction with objects. And elastomers find use as seals when dust or liquids need to be kept out or in.

Rubber

The word rubber is often used in a generic sense to refer to any elastomeric material. More specifically, the term refers to carbon-backbone polymeric elastomers, and even more specifically, to those derived from the natural material. Natural rubber is made from a milky sap called latex produced by a variety of plants. The commercial source is the tropical rubber tree (Hevea brasiliensis), a kind of euphorbia, but many other species produce rubbery latex including milkweed and dandelions. Its natural role appears to be to permanently gum up the mouthparts of insects that would eat the plant. Exposed to air, latex polymerizes into a soft, resilient material that can be heated and remolded. Charles Goodyear discovered that when heated with a bit of sulfur, latex becomes much firmer, stronger, and abrasion resistant. He called the process vulcanization and the tire industry was born. Chemically, the sulfur crosslinks individual polymer chains making the material more rigid and resilient.

Natural rubber is still used to make premium tires, especially for trucks. But during and after WWII, demand for rubber exceeded the natural supply leading to intense research into synthetic alternatives. Today we have dozens, with all sorts of special properties. The first commercial synthetic success, still widely used, was based on polybutadiene (butane with two double bonds). The second double bond both makes the chains curl into springy coils and provides opportunities for cross-linking.

Rubber can be obtained pre-formed in a variety of shapes. Commonly available forms include O-rings, corks, sheet material, and rubber bands. Big, specially-shaped chunks show up as isolation mounts in a variety of machinery. Sheet material can be cut with scissors, and there are special tools for cutting round holes in corks or other solid forms, but otherwise, cured rubber is difficult to shape. It glues well with special cements, and can be attached to rigid substrates with superglue. Because superglue is rigid, it does not work well for rubber-to-rubber contacts that will be stretched.

Silicone Rubber

Silicone rubbers are elastic polymers having a silicon-containing backbone (instead of or in addition to carbon). Note that these polymers are distinct from elemental silicon, a brittle, silvery material from which computer chips are made. Some silicones are available in an uncured form that sets on exposure to air to a rubbery solid (e.g. "silicon-seal"). These materials have a variety of uses in robot construction. They are ideal for making gaskets in place, and for making joints between assembled components waterproof or even airtight. They can also be used to "glue" certain materials to each other that cannot normally be glued, in particular glass and metal. The bonds are not strong in an absolute sense, but they are quite secure for holding components in place against modest forces since the flexibility prevents fracture failure.

Flexible Tensile Elements

Tensile elements can be considered a distinct material class in the same sense as rigid solids and elastomers. They have strength only in tension in one dimension, and are otherwise flexible. The most familiar form is multi-strand twisted or braided cable, but solid wires are occasionally used. Chains and flexible belts are more specialized examples.

Structural cables are made from relatively few materials, the most important being stranded steel wire (also known as wire rope), and a handful of polymer fibers including polyester (Dacron), aramid (Kevlar), and UHMWPE (Spectra, Dyneema). Nylon and polypropylene are frequently used for rope, but they are too stretchy for most structural applications. Chains are usually made of steel links, and belts from a combination of tensile fibers with elastomers.

Tension elements find two primary uses in robotics. The first is simply as tensile reinforcement in rigid structures. A rectangular frame is mechanically weak and prone to failure by diagonal skewing. Reinforced by a pair of diagonal cables, the structure is strong and rigid in two dimensions, and very light. In three dimensions, there is still a racking (twisting) failure mode, but rigid 3-D structures can be made by combining such 2-D panels. The other use is in transmissions - as tendons, belts, or chains. Transmissions are a separate subject, but some of the simplest and best depend on the existence of flexible tension elements.

References

- Robotics, design: galileo.org, Retrieved 11, May 2020
- Robotics-software-tools: digitalengineering247.com, Retrieved 13, June 2020
- Electronics-basic-components-tutorial: societyofrobots.com, Retrieved 27, July 2020
- What-are-the-primary-mechanical-components-of-a-robot: robotsauthority.com, Retrieved 20, April 2020
- Materials, robot-manual, csc-robocon, nelson: cs.rochester.edu, Retrieved 25, August 2020

4

Robotic Parts

Robots are made up of many components. The key components of robots are actuator, end-effector, manipulator, servomechanism, stepper motor, and robotic sensors etc. This chapter has been carefully written to provide an easy understanding of the various components of the robots.

Actuator

Actuators have become a crucial part of the automated system, as they help with controlling equipment using hydraulic, pneumatic or in some of the cases electronic signals. Linear actuators are quite common as they form the bulk of applications used in the control of equipment moving along a lateral axis. However it is also important to note that there are other types of actuators that are found in robotic arms. One of the actuators types is referred to as a synchronous actuator and relates to the fact that it has a motor which rotates synchronously with the oscillating current. The brushless DC servo is also a common one and like the name suggests it uses a DC current and does not have brushes.

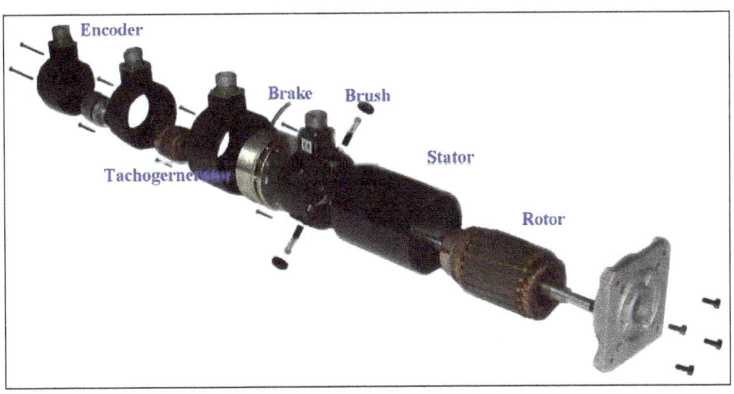

Instead it has permanent magnet poles on the rotor which are attracted to the magnetism of the rotating poles. They tend to last longer and give off no noise along with reduced interference from electromagnetism.

The stepper is also brushless and moves in small but discrete steps. It can be used with digital and analog signals. The brushed DC servo on other hand has brushes/commutators which are found on a permanent magnet or a wound stator. The asynchronous actuator on the other hand is designed to slip for it to generate torque. Traction motor is an electric motor that is common cars for driving the wheels. AC servo motors are two phased just like the driving current and are crucial where fast and accurate responses are required.

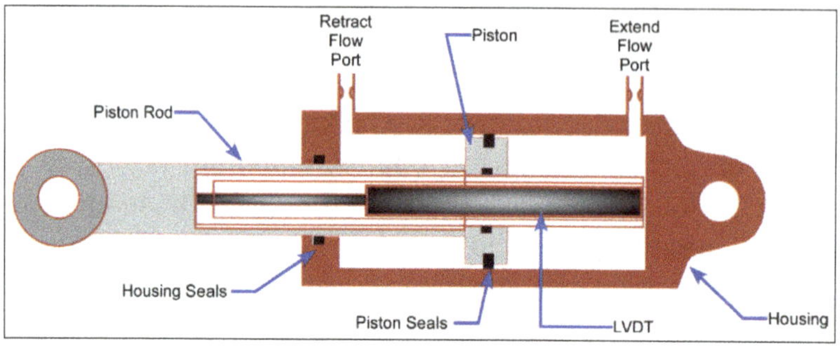

Types of Robot Actuators

Pneumatic Actuator

These actuators use the principle of Pneumatics (where pressurized air/gas) to create motion (or perform work). Pneumatic actuators can be used to produce both rotational motion and linear motion. Pneumatic Actuator The basic design includes a Cylinder, a Piston and few supporting components like valves, springs, stoppers etc., Compressed Air or pressurized gas is filled into the cylinder and the compressed air tries to expand to reach atmospheric pressure. This expansion is forced towards a piston or any other mechanical device which makes an attached object move. Depending on the design, it can either create rotational motion or linear motion. Most air compressors and air pumps use this principle due to its simplistic design. Image shows single acting cylinder where a spring pushes the piston back to its base position. In a double acting cylinder, another valve on other end of the cylinder pushes the piston back to base position.

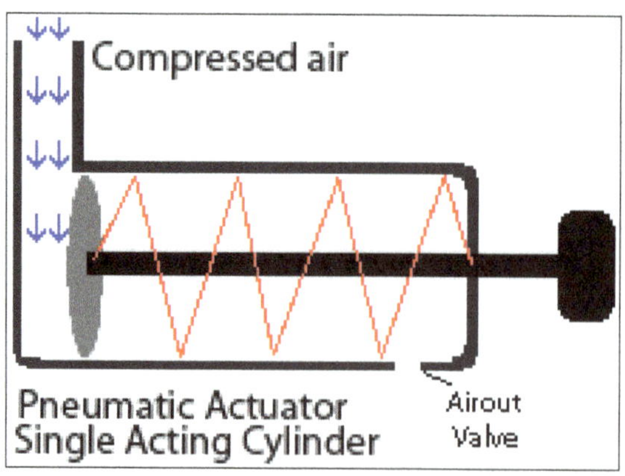

Pneumatic actuators are mainly used for systems which require quick and accurate response. These actuators are clean, make less noise and relatively compact in their design.

Air Muscles: To replicate muscles in robots, Air muscles (also known as pneumatic air muscles - PAM) are used. PAM's generally consist of a rubber bladder covered by a braided fiber mesh. When pressurized gas/air is inflated, it expands radically and when deflated, it contracts. Air muscles are inexpensive, light weight, exhibit a phenomenal strength to weight ratio, easier to build, flexible compared to other electric and hydraulic actuators, durable, safe, and also easy to use under water.

Hydraulic Actuator

According to Blaise Pascal, when there is an increase in pressure at any point in a confined incompressible fluid, then there is an equal increase at every point in the container. Hydraulic actuators are designed based on this principle (Pascal's law).

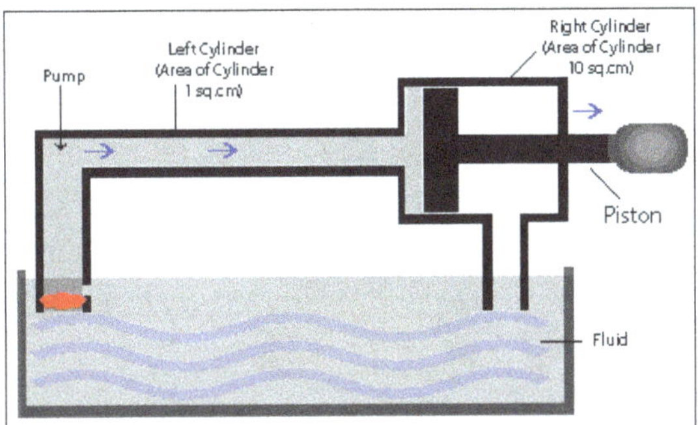

Hydraulic Actuator

To understand how hydraulic actuators work, let us take an example of two cylinders, connected together, as shown in the image. Suppose one cylinder has a cross-section area of 1 square centimeter and the second one has a cross-section area of 10 square centimeters. If the cylinders are filled with incompressible fluid and 1 unit of pressure is applied to the left cylinder pushing the pump (actually liquid) by 10 centimeter, then the resulting force acts on the right cylinder pushing the piston by 1 centimeter, but with a force of 10 units. This means applying 1 unit of force produces 10 units of force on the other side.

Hydraulic actuators are majorly used for systems which require very large force, but not very restrictive on positioning and accuracy.

Hydraulic Cylinder: Generally referred as linear hydraulic motor is used when a robot requires linear force. These actuators are powered by hydraulic fluids. When hydraulic pressure acts on it, a piston connected to a piston rod within the cylinder moves back and forth creating linear motion. Hydraulic cylinders can also convert hydraulic pressure into rotation creating a rotary actuator, based on the mechanical design.

Piezoelectric Actuators

These actuators use piezoelectric effect to create motion. For those who are not aware of piezoelectric (piezo means press, or apply pressure) effect, it is a charge that accumulates in certain ceramic materials and crystals (like quartz, tourmaline etc.) when a mechanical stress is applied. This is known as piezoelectric effect where deformation of the material creates a voltage difference.

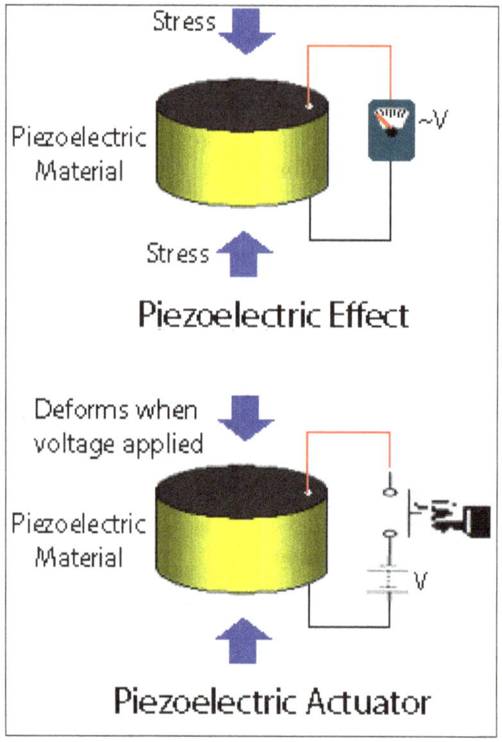

Piezoelectric Effect Piezoelectric Actuator.

When an electricity flows through a piezoelectric material, it creates a physical deformation which is proportional to the applied electric field, known as indirect piezoelectric effect. This precise deformation can be used to position objects with extreme accuracy, almost at μm accuracy.

Piezoelectric actuators are used in Loudspeakers, Piezoelectric Motors, acceleration sensors, vibration sensors etc., and can be used to create either rotational or linear motion. The strokes of these actuators can also be amplified if required, because direct strokes from these peizoelectric actuators are generally less than 100 μm.

Ultrasonic Piezoelectric Actuators

Similar to Piezoelectric actuators, Ultrasonic Piezoelectric actuators work on the principle of piezoelectric effect. A travelling wave excites a stator surface (material exhibiting piezoelectric effect) which behaves like an elastic ring and produces elliptical motion at the interface of the rotor, which propels the rotor and the drive shaft connected to it.

Ultrasonic Piezoelectric actuator can induce either rotational motion or linear motion.

These actuators provide extremely precise movement with good torque to size ratio, and used in most camera autofocus lenses and watch motors.

Delta Robot

Delta robots are a type of parallel robot, meaning that their motors and legs are connected mechanically in parallel, with each leg supporting only itself. In contrast, SCARA, 6-axis, and Cartesian styles are serial robots, meaning that each arm has to carry the next arm beyond it. For example, in an X-Y-Z Cartesian robot, the Y axis must support the Z axis, and the X axis must support both the Y and Z axes.

Delta robots operate on the basis of three parallelograms connected to a stationary base above and a moving platform below. The base is essentially an equilateral triangle, with three legs connected to it—one on each side of the triangle—each moved by a small motor. These legs are jointed (think of knees), with the lower sections consisting of two bars in parallel, and the lower end of each bar connected to the moving platform, which is also an equilateral triangle. These lower sections of the legs make up the three parallelograms that control the platform, constraining it to X, Y, and Z motion, parallel to the base.

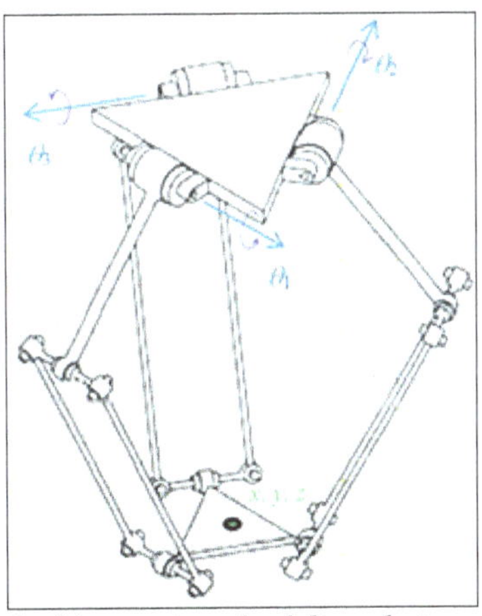

From a mathematics standpoint, the kinematics of a delta robot are relatively straightforward. Three motors manipulate three parallelograms to move a platform in the X, Y, and Z directions.

A fourth axis can be included for rotation, either by attaching a rotary actuator directly to the platform, or by adding a leg that extends from the center of the base to the center of the platform, with a rotating joint on the end. (Note that this leg must be telescoping, as its length will change as the three primary legs change position).

Design Advantages of Delta Robots

Delta robots dominate the field when it comes to speed and acceleration, because only their legs and platform move (along with the payload, of course). This means that they have much less moving mass and inertia than SCARA, 6-axis, or even Cartesian robots, and can achieve accelerations in the range of 150 m/s2 and maximum speeds up to 10 m/s. To further capitalize on this design benefit, the legs and platform can be made from lightweight materials, such as aluminum or carbon fiber.

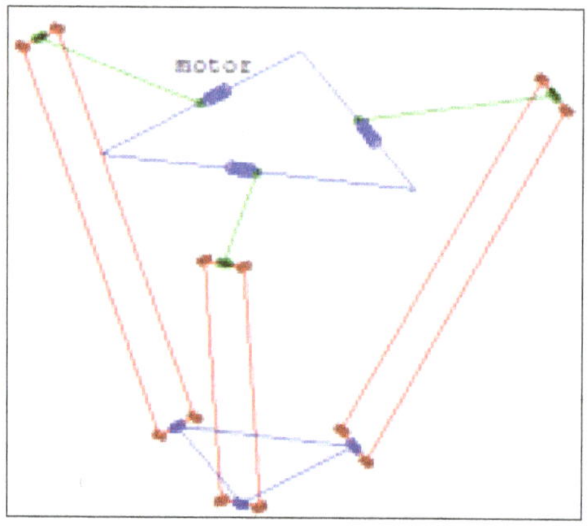

Three parallelograms (in red), are connected to the stationary base and moving platform (in blue).

Unlike serial robots, in which the errors in each arm accumulate, in the parallel robot design, the errors in each leg are averaged. This gives delta robots the ability to achieve sub-millimeter repeatability. The parallel design also provides high stiffness, because the legs work together against external forces and torques. (Note that the stiffness of the legs, which is influenced by their material and size, also has an effect on the robot's stiffness).

Being mounted above the workspace, delta robots have no interferences in their work area, meaning there are no "dead zones," or zones that the end effector cannot reach within the defined work envelope. Overhead mounting also makes delta robots easy to adapt for harsh or challenging environments. With the major electromechanical parts mounted above the work area, they can be shielded from harsh chemicals or debris, or enclosed to make them clean-room or wash-down compatible. And the legs can be made of virtually any material that is suitable for the environment.

Applications Then and Now

Delta robots were introduced for industrial applications in the early 1990's, with the first robots being employed in the packaging industry. As vision systems and force sensors were integrated, delta robots began to move beyond simple pick-and-place operations, and by the late 1990's, their applications had expanded into the electronics and assembly markets.

More recently, delta robots are being deployed in high-speed assembly and light machining applications. And new industries, such as 3D printing, are taking advantage of their relative mechanical simplicity and high-speed capabilities. Now, delta robots can be found in virtually any application that requires the ability to achieve extremely fast speeds and accelerations with light payloads.

End Effector

An End Effector is considered as the "hand of a robot." It is one of the important devices in a robot. It integrates an arm and a wrist, which helps it to perform several functions like material handling, pick and place, machine loading and unloading, etc. It is also known as EOA (End of Arm Tooling), Robotic Peripherals, and more.

Based on the performance of work tasks, the robot end effector can be described into two important categories such as:

- Grippers, and
- Tools.

Grippers

The grippers are mainly used in a robot for grasping / holding an object. The grasped objects will be moved to the preferred place with the help of a robot. The grippers are capable of carrying work parts, bottles, tools, and so on. It can work by means of magnetic, mechanical, vacuum cups, etc.

The grippers can also be sorted as:

- Single gripper,
- Double grippers, and
- Multiple grippers.

If a robot wrist contains one grasping device, then it is described as Single Gripper. Similarly, a robot wrist with two grasping devices will be termed as Double Grippers. It will be very much suitable for machine loading and unloading operations. Moreover, it is better than single gripper, because it finishes plenty of works in a quick time. The Multiple Grippers incorporates more than two grasping devices in a robot wrist. The double grippers seem to be the best one than the multiple grippers. The reason is that the multiple grasping mechanisms are pricing very high. The grippers can be further classified into internal gripper and external gripper.

Other types of grippers are:

- Magnetic Grippers,

- Vacuum Cups, and
- Adhesive Grippers.

A tool is equipped in the robot for carrying out several operations on the work parts instead of grasping it. A tool acts as an end effector when it is attached directly to the robot's wrist. In some applications, there will be a need for multi-tool task, and changing the tool all the time from the robot wrist will be highly difficult. As a result, a gripper is used in this process to grasp and manipulate the tool. It certainly helps the robot to handle several tools in an operation, and thus makes the multi-tooling function possible. Moreover, the time taken to change the tool is very low.

For example: In a deburring operation, various sizes of deburring tool will be required to hold for reaching every surface of the work part. Here, the tool is equipped in the gripper for quick exchange from one tool to another.

Examples for Tools used as end Effectors

In robot applications, the most commonly used three tools as end effectors are listed below:

- Spot welding tools.
- Spray painting nozzle.
- Arc welding torch.

The rotating spindle for routing, drilling, grinding, and wire brushing operations also comes under this category. Some other examples of tools used as end effectors are liquid cement applicators, water jet cutting tool, and heating torches.

In all the above examples, the actuation of a tool must be coordinated by an industrial robot. An example of this case includes the spot welding operation in which the robot must control the actuation as a part of its process. The control of this process is very similar to the opening and closing of a mechanical gripper.

Forward Chaining

Often rule-based systems work from just a few facts but are capable of reaching many possible conclusions. Examples are "sensory" and "diagnosis" expert systems, like those that identify an object from a description of what you see at a distance, or those that tell you what to do when your car breaks down from a description of what isn't working. For these, it often makes more sense to start with the facts and reason to goals, what is known as forward chaining or data-directed computation or modus ponens reasoning.

As an example, take the rule:

a :- b, c.

and suppose b and c are facts. Then we can conclude a is a fact, and add it to the facts we have. This is called modus ponens inference in logic. There's no query, no goals; we just use the facts we know to infer some new fact. (The built-in predicate asserta can help implement this too).

To use modus ponens as the basis of a control structure, take the facts in order. For each fact, find all rules whose right sides contain a predicate expression that can be matched to it. (We can index predicate names mentioned on right sides to speed this up.) Now "cross out" the predicate expressions matched in the rules; that is, create new rules like the old rules except without these expressions. But wherever a fact matched the last expression on the right side of some rule, the left side of the rule has been proved a new fact, after substituting in any bindings made on the right side; so cache that new fact in the database. The last paragraph gave an example. For another, consider:

a(X,3) :- b(X).

Then if b(20) is a fact, modus ponens would conclude a(20,3) is a fact.

It matters where we put a new fact among the others, since the facts are pursued in order. Usually it's good to put the new fact in front of those not yet considered, a most-recent-fact or focus-of-attention idea. So the fact just proved will be the next fact examined; the fact list then is much like a stack (last thing "in" is the first thing "out"). This might be good if we want to reach some particular conclusion as fast as possible. But, on the other hand, if we want to systematically find every conclusion possible from the facts, we should put new conclusions at the end of the set of facts; the fact list is then like a queue (last thing "in" is the last thing "out").

Any nots in rules require special handling. Since we want to follow the closed-world assumption for forward chaining too (it's simplest), we want not to mean "it can't be proved". So forward chaining must first assume all nots to be false, prove all possible facts, and only then consider as true those nots whose arguments are not now facts. Those nots may then prove new facts with new consequences. (To avoid such awkwardness, some artificial-intelligence systems let you state certain facts to be false, but this creates its own complications).

Here's the formal algorithm for (pure) forward chaining:

1. Mark all facts as unused.

2. Until no more unused facts remain, pick the first-listed one; call it F. "Pursue" it:

    (a) For each rule R that can match F with a predicate expression on its right side, ignoring nots, and for each such match in the rule (there can be multiple match locations when variables are involved):

(i) Create a new rule just like R except with the expression matching F removed. If variables had to be bound to make the match, substitute these bindings for the bound variables in the rule.

(ii) If you've now removed the entire right side of rule R, you've proved a fact: the current left side. Enter that left side into the list of facts, and mark it "unused". (The focus-of-attention approach here puts the new fact in front of other unused facts.) Eliminate from further consideration all rules whose left sides are equivalent to (not just match able to) the fact proved.

(iii)Otherwise, if there's still some right side remaining, put the new simplified rule in front of the old rule. Cross out the old rule if it is now redundant. It is redundant if the old rule always succeeds whenever the new rule succeeds, which is true when no variables were bound to make the match.

(b) Mark F as "used".

3. Create a new fact list consisting of every not expression mentioned in rules whose argument does not match any used fact. Mark all these as "unused", and redo step 2.

A Forward Chaining Example

Let's take an example of forward chaining with focus-of-attention handling of new facts.

```
/* R1 */ goal1 :- fact1

/* R2 */ goal1 :- a, b

/* R3 */ goal2(X) :- c(X)

/* R4 */ a :- not(d)

/* R5 */ b :- d

/* R6 */ b :- e

/* R7 */ c(2) :- not(e)

/* R8 */ d :- fact2, fact3

/* R9 */ e :- fact2, fact4
```

Suppose the rules are taken in the given order; and that as before, only fact2 and fact3 are true, in that order.

1. We start with fact2, and find the matching predicate expressions R8 and R9. This gives the new rules:

```
/* R10 */ d :- fact3
/* R11 */ e :- fact4
```

Rules R8 and R9 are now redundant since no variables were bound, and R8 and R9 can be eliminated.

2. No new facts were discovered, so we pursue next fact3. This matches an expression in R10. So now R10 succeeds, and the new fact d is put in front of any remaining unused facts (though there aren't any now). R10 can be eliminated.

3. We pursue fact d, and find that rule R5 mentions it in its right side. (R4 mentions it too, but as not(d), and we're saving nots for last.) Matching R5 gives the new fact b. Rules R5 and R6 can now be eliminated.

4. Fact b matches in R2, giving.

```
/* R12 */ goal1 :- a
```

and rule R2 can be eliminated. The current set of rules is:

```
/* R1 */ goal1 :- fact1
/* R3 */ goal2(X) :- c(X)
/* R4 */ a :- not(d)
/* R7 */ c(2) :- not(e)
/* R11 */ e :- fact4
/* R12 */ goal1 :- a
```

5. We have no more facts to pursue. But we're not done yet, since R4 and R7 have nots.

6. Fact d is true, so rule R4 can't ever succeed. But fact e has not been proved. Hence add not(e) to the list of facts.

7. This matches the right side of R7. Hence c(2) is a fact too. Eliminate R7.

8. This matches the only expression on the right side of R3, when X=2, and hence goal2(2) is a fact. We can't eliminate R3 now because we had to bind a variable to make the match, and we can still use R3 for other values of X.

9. That's everything we can conclude.

Manipulator

Industry-specific robots perform several tasks such as picking and placing objects, movement adapted from observing how similar manual tasks are handled by a fully-functioning human arm. Such robotic arms are also known as robotic manipulators. These manipulators were originally used for applications with respect to bio-hazardous or radioactive materials or for use in inaccessible places.

A series of sliding or jointed segments are put together to form an arm-like manipulator that is capable of automatically moving objects within a given number of degrees of freedom. Every commercial robot manipulator includes a controller and a manipulator arm. The performance of the manipulator depends on its speed, payload weight and precision. However, the reach of its end-effectors, the overall working space and the orientation of the work is determined by the structure of the manipulator.

Kinematics of a Robotic Manipulator

A robot manipulator is constructed using rigid links connected by joints with one fixed end and one free end to perform a given task (e.g., to move a box from one location to the next). The joints to this robotic manipulator are the movable components, which enables relative motion between the adjoining links. There are also two linear joints to this robotic manipulator that ensure non-rotational motion between the links, and three rotary type joints that ensure relative rotational motion between the adjacent links.

The manipulator can be divided into two parts, each having different functions:

Arm and Body – The arm and body of the robot consists of three joints connected together by large links. They can be used to move and place objects or tools within the work space.

Wrist – The function of the wrist is to arrange the objects or tools at the work space. The structural characteristics of the robotic wrist includes two or three compact joints.

Robotic Manipulator Arm Configuration

- Cartesian geometry arm – This arm employs prismatic joints to reach any position within its rectangular workspace by using Cartesian motions of the links.
- Cylindrical geometry arm – This arm is formed by the replacement of the waist joint of the Cartesian arm with a revolute joint. It can be extended to any point within its cylindrical workspace by using a combination of translation and rotation.
- Polar/spherical geometry arm – When a shoulder joint of the Cartesian arm is

replaced by a revolute joint, a polar geometry arm is formed. The positions of end-effectors of this arm are described using polar coordinates.

- Articulated/revolute geometry arm - Replacing the elbow joint of the Cartesian arm with the revolute joint forms an articulated arm that works in a complex thick-walled spherical shell.
- Selective compliance automatic robot arm (SCARA) – This arm has two revolute joints in a horizontal plane, which allow the arm to extend within a horizontal planar workspace. The TH650A SCARA Robot by TM Robotics is a great example to demonstrate pick and place functionality of robotic manipulators.

Wrist Configuration

The two main types of wrist design include:

- Roll-pitch-roll or spherical wrist.
- Pitch-yaw-roll.

The spherical wrist is more common because of its mechanically simpler design. It has 6 degrees of freedom, and consists of a Hooke shoulder joint followed by a rotary elbow joint.

Applications

Some of the major applications of robotic manipulator are discussed below:

- Motion planning,
- Remote handling,
- Micro-robots,
- Humanoid robots,
- Machine tools.

Parallel Manipulator

Parallel manipulators are widely popular recently even though conventional serial manipulators possess large workspace and dexterous maneuverability. The basic problems with serial one are their cantilever structure makes them susceptible to bending at high load and vibration at high speed leading to lack of precision and many other problems. The serial one are compared. Hence, in applications demanding high load carrying capacity and precise positioning, the parallel manipulators are the better alternatives and the last two decades points to the potential embedded in this structure that has not yet been fully exploited. Willard L. V. Polard designed and patented the first industrial parallel robot as shown in. The development of parallel manipulators can be dated back to the early 1960s, when Gough and Whitehall first devised a six-linear jack system for use as a universal tire testing machine. Later, Stewart developed a platform manipulator for use as a flight simulator. Since 1980, there has been an increasing interest in the development of parallel manipulators. This paper highlights the potential applications of parallel manipulators include mining machines, walking machines, both terrestrial and space applications including areas such as high speed manipulation, material handling, motion platforms, machine tools, medical fields, planetary exploration, satellite antennas, haptic devices, vehicle suspensions, variable-geometry trusses, cable-actuated cameras, and telescope positioning systems and pointing devices. More recently, they have been used in the development of high precision machine tools by many companies such as Giddings & Lewis, Ingersoll, Hexel, Geodetic and Toyoda, and others. The Hexapod machine tool [5,6] is one of the widely used parallel manipulators for various industries.

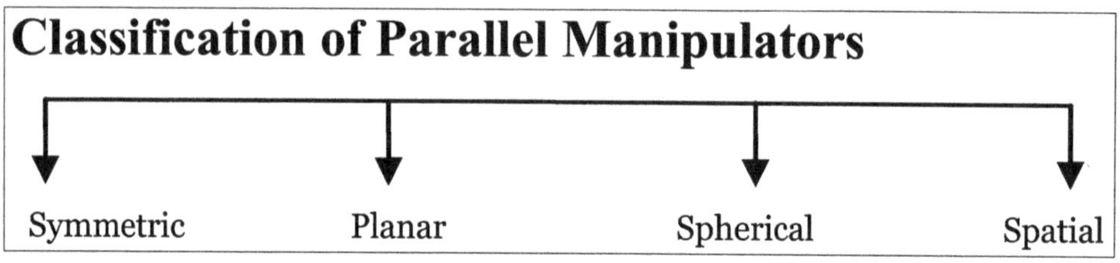

Symmetrical manipulators has number of limbs equals to number of degree of freedom, which is also equals to total numbers of loops. A planar parallel manipulator is formed when two or more planar kinematic chains act together on a common rigid platform. Now days, each leg of a planar parallel manipulator is replaced by a single wire, the manipulator is referred to as a planar wire-actuated (or wire-suspended) parallel manipulator. Spherical manipulators are just able to make the end effectors movement according to controlled spherical motions. Especially, spatial parallel manipulators with fewer degrees of motion than six, but more than three, attracted the attention of both, the researchers and users.

In the most of the research work, kinematic analysis is carried by either kinematic constraint equations, screw theory method, DH Parameters or using the concept of dual unit quaternion which is really a most efficient for representing screw displacements of lines for spatial manipulators. Using the inverse dynamics, the forces and torques of the actuated joints can be computed for closed loops configurations of parallel architecture. Newton Euler approach is relatively more economical compared to Euler Lagrange formulation for parallel as well as hybrid manipulators.

Industrial Applications

In 1942, a patent was issued to Willard L. V. pollard for his novel design of automatic spray painting. In fact, it was never built. It has five degree-of-freedom motion—three for the position of the tool head, and the other two for orientation. In 1954, Dr. Eric Gough, an employee of Donlop Rubber Co., England had developed a six degree of freedom—the first octahedral hexapod for universal Tyre Testing Machine. In 1965, Stewart published a paper in which he proposed a six degree of freedom parallel platform as a flight simulator. The Stewart platform comprises six pods, six spherical joints and six universal joints with 6 degree of freedom as shown in. Stewart platform has also been used for the Agile Eye (a spherical Parallel mechanism) —pointing devices developed by Gosselin and Hamel at robotics laboratory at Laval University, Canada, has low inertia and inherent stiffness, the mechanism can achieve angular velocities superior to 1000 deg/sec and angular accelerations greater than 20,000 deg/sec2, large- ly outperforming the human eye. An American engineer, Klaus Cappel is considered as a third pioneer in field of parallel robotics. He has developed an octahedral Hexa- pod Manipulator as motion simulator & was patented in 1967 as shown in F. Stewart Platform is also used for underground excavation device by Arai in 1991 and another application such as milling machines by Aronson in 1996.

3-DOF spherical parallel mechanism as shown in has interesting characteristics for practical ap- plications like worktable, a manipulator, a camera ori- enting device, a wrist or a motion simulator. Parallel Robots can offer many advantages for high-speed laser operations as shown due to their structural stiffness and limited moving masses with less power consumption.

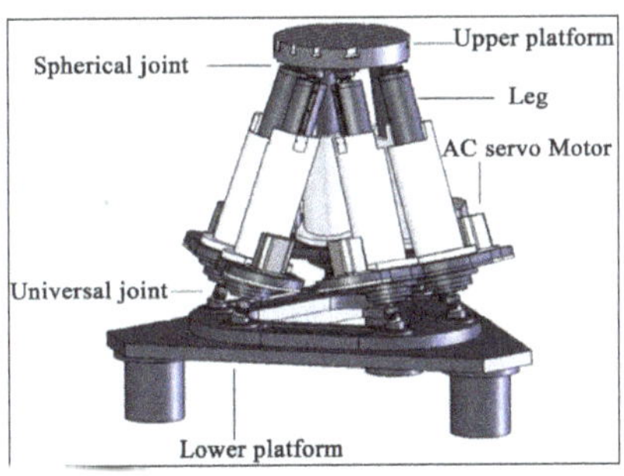

Typical Gough Stewart platform.

First design of an octahedral hexapod for flight simulator (courtesy of Klaus Cappel).

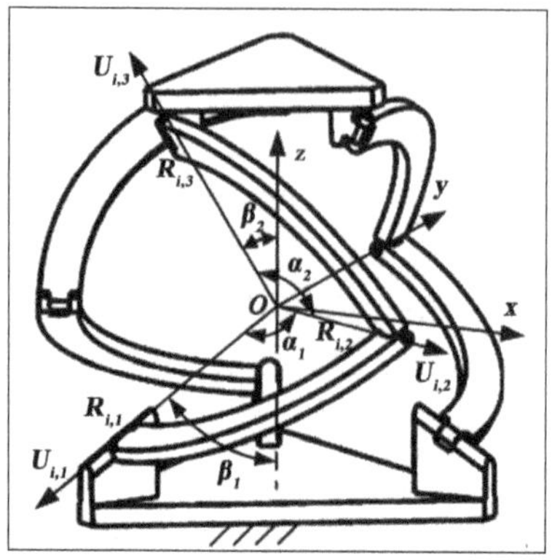

Three DOF spherical parallel manipulators.

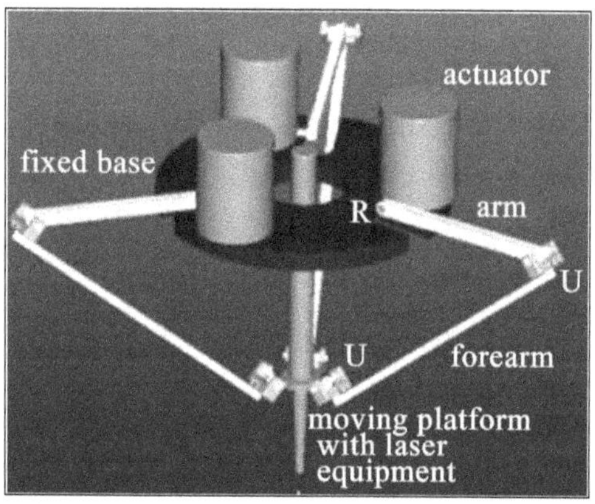

Three RUU kinematic chains usage in delta robot for laser cutting.

depicts the parallel cube-manipulators have characteristics of no singularities in the workspace, simple form of forward kinematics and existence of a compliance center x = y = z = 0 with a added advantages of high stiffness and compactness used in the fields of micro-motion manipulators, remote center compliance (RCC) devices, assembly, planar kinematics machines and so on.

Remote Manipulator

Several stations have built in features that allow equipment, such as TRIUMPH cameras, to be deployed to specific locations in the reactor. These machines used to carry out this deployment are called manipulators and come in many shapes and sizes.

The size and shape of the manipulators is generally determine by the need to ensure that reactor gases do not escape whilst carrying out ISI – this is known as keeping containment. In order to achieve this most manipulators have a containment box or a sealing system and when in use are sealed onto the reactor penetration.

Manipulators are generally controlled from a dedicated console which contains joysticks, drivers, power supplies, camera monitors, computers and indicators. The power and signals to and from the manipulator are carried via umbilical which can be heavy and difficult to handle.

Example of Remote Manipulator.

1. Hoist Unit

This is a very simple form of manipulator and basically consists of a winch housed inside a containment box. In order to keep containment, operations are carried out within the containment box which have built in gloves. These holes are known as gloveports. They can also be sealed off more permanently by use of gloveport valves or seals.

Hoist units are generally used to lower and raise equipment vertically in and out of the reactor core. These are generally used to inspect fuel and control rod channels as they can be easily positioned vertically above the channel.

To inspect components which are not vertically below the manipulator, a set of arms, similar to a crane davit, can be fitted inside the containment and used to lift the equipment sideways. From this new position the equipment can be lowered and raised. These arms are known as Elevating Guide Tubes (EGT) and can typically reach up to 2m.

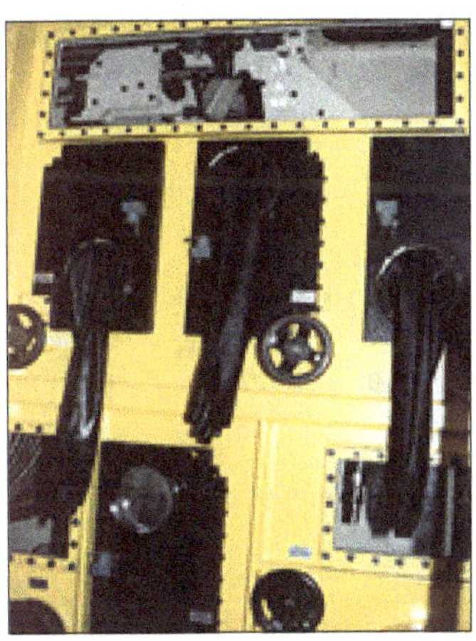

Gloveport.

2. Links Manipulators

In the majority of cases EGTs have insufficient reach to position the camera in the desired location. In these situations the cameras are deployed by links manipulators which have variable extensions up to 7m. The rigid extending arm is formed by a series of links, typically made from titanium. These are hinged at the top and can only bend in one direction. When deployed horizontally the hinges bend until the faces of adjoining links butt up against each other forming the arm. As the penetrations into the reactor are generally vertical, it is necessary to insert the links vertical then bend them around to the horizontal.

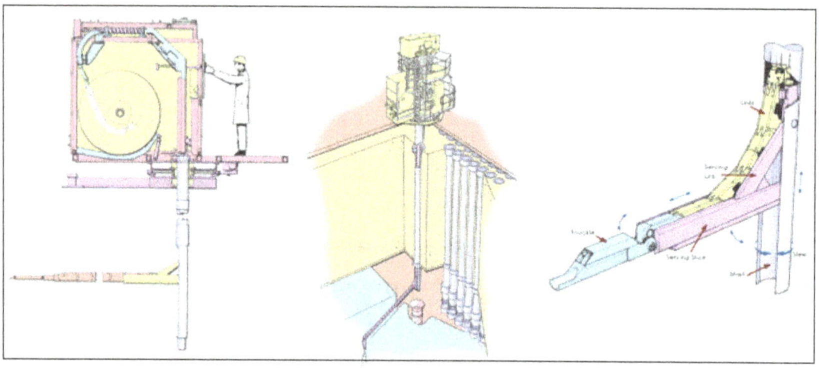

Examples of how a link manipulator functions.

3. Extending Arm Manipulator

Another type of machine currently in use is the extending arm manipulator. In these devices the rigid arm is formed by extending telescopic sections. These manipulators are particularly useful where access can be gained horizontally, for example via the gas circulator penetrations.

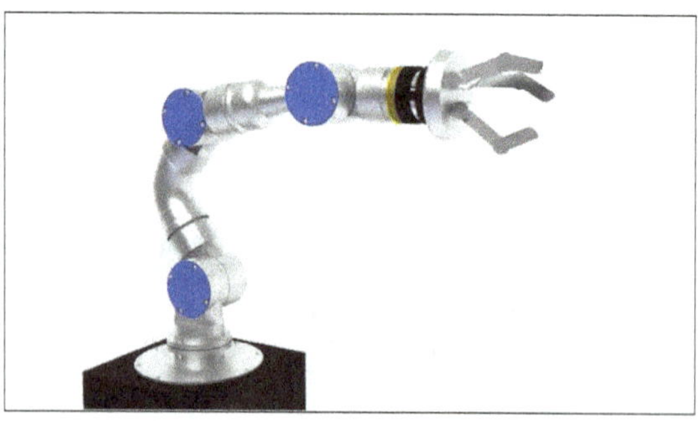

Example of an extending arm manipulator.

Serial Manipulator

A serial manipulator consists of a fixed base, a series of links connected by joints, and ending at a free end carrying the tool or the end-effector. In contrast to parallel manipulators, there are no closed loops. By actuating the joints, one can position and orient the end-effector in a plane or in three-dimensional (3D) space to perform desired tasks with the end-effector. This chapter deals with kinematics of serial manipulators where the motion of links are studied without considering the external forces and torques which cause these motions. The serial manipulator geometries are described using the well-known Denavit-Hartenberg (D-H) parameters. Two well-known problems, namely the direct and inverse kinematics problems, are posed, their solution procedures discussed in detail and illustrated with examples of planar and spatial serial manipulators. It is shown that closed-form analytic solutions to the inverse kinematics problem is possible only for serial manipulators with special geometries and the most general six-degree-of-freedom serial manipulator requires the solution of at most a 16th degree polynomial. The solution of the inverse kinematics problem leads to an important and useful concept of the workspace of a serial manipulator and the approaches to obtain the workspace and determine its properties are also presented.

Degrees of Freedom of a Manipulator

The degrees of freedom of a serial manipulator1 can be obtained from the well-known Chebychev- Grübler-Kutzbach criterion:

$$\text{dof} = \lambda(N - J - 1) + \sum_{i=1}^{j} Fi$$

where dof is the computed degree of freedom with N as the total number of links including the fixed link (or base), J as the total number of joints connecting two consecutive links, Fi as the degrees of freedom at the ith joint,

$\lambda = 6$, for motion in $3D$

3, for planar motion

The quantity, dof, obtained from equation dof $= \lambda(N - J - 1) + \sum_{i=1}^{j} Fi$ is the number of independent actuators that can be present in the serial manipulator. In a broad sense, dof determines the capability of the serial manipulator with respect to dimension of the ambient space λ. We have the following possibilities:

1. dof $= \lambda$ – In this case, an end-effector of a manipulator can be positioned and oriented arbitrarily in the ambient space of motion.
2. dof $< \lambda$ – In this case, the arbitrary position and orientation of the end-effector is not achievable and there exist $(\lambda$ – dof$)$ functional relationships containing the position and orientation variables of the end-effector.
3. dof $> \lambda$ – These are called redundant manipulators and the end-effector can be positioned and oriented in infinite number of ways.

In serial manipulators with a fixed base, a free end-effector and two links connected by a joint, from equation $N = J + 1$ and dof $= \sum_{i=1}^{J} Fi$. If all the actuated joints are one-degree-offreedom joints, then $J = \text{dof}$.

If $J < dof$, then one or more of the actuated joints are multi- degree-of-freedom joints and this is not used in mechanical serial manipulators. This is due to the fact that it is difficult to locate and actuate two or more one- degree-of-freedom joints at the same place in a serial manipulator. In biological systems, muscles are used to actuate multi- degree-of-freedom joints – in a human arm muscles actuate the three- degree-of-freedom shoulder joint.

In manipulators, the J joint variables form the joint space. The variables describing the position and orientation of a link or the end-effector are called the task space variables. The dimension of task space is ≤ 6 for 3D motions and ≤ 3 for planar motion. Finally, there are often mechanical linkages, gears, etc. between actuators and joints. The space of all actuator variables is called the actuator space. If the dimension of the actuator space is more than 3 for planar motion and more than 6 for 3D motion, the manipulator is called redundant. If the dimension of the actuator space is less than the degree of freedom, then the manipulator is called under-actuated.

Servomechanism

Servomechanism refers to a device or combination of devices that automatically controls a mechanism or a source of power or energy. Servomechanisms automatically compare the controlled output of a mechanism to the controlling input. The difference between the settings or positions of the output and the input is called the error signal, which regulates the output to a desired value. Servomechanisms may be mechanical, electrical, hydraulic, or optical. The process of sending the error signal back for comparison with the input is called feedback, and the whole process of the input, output, error signal, and feedback is called a closed loop. The closed-loop system, also known as a servomechanism, has some means of incorporating mechanical feedback from the output to the input. A sensor at the output end generates a signal that is sent back to the input to regulate the machine behavior. The term servomechanism correctly applies only to systems where feedback or error-correction signals help control mechanical position or other parameters. For example, an automotive power window control is not a servomechanism, because there is no automatic feedback which controls position—the operator does this by observation. By contrast the car's cruise control uses closed loop feedback, which classifies it as a servomechanism.

Purpose of Servomechanism:

1. Accurate control of motion without the need for human attendants (automatic control),

2. Maintenance of accuracy with mechanical load variations, changes in the environment, power supply fluctuations, and aging and deterioration of components (regulation and self- calibration),

3. Control of a high-power load from a low-power command signal (power amplification) and,

4. Control of an output from a remotely located input, without the use of mechanical linkages.

A servomechanism is unique from other control systems because it controls a parameter by commanding the time-based derivative of that parameter. For example a servomechanism controlling position must be capable of changing the velocity of the system because the time-based derivative (rate change) of position is velocity. A hydraulic actuator controlled by a spool valve and a position sensor is a good example because the velocity of the actuator is proportional to the error signal of the position sensor. A simple example is the driver of a car. His eyes tell him where he is on the road, and compare it to where he should be, and this information makes its way to his brain. The brain decides what action should be taken in order to move the car from where it is to where it should be, and sends a signal to the muscles in the arm, turning the steering wheel to realign the car.

All servomechanisms have the following parts:

1. A way to measure what is desired and what is being accomplished,

2. A way to transport this information,

3. A way to determine the difference between the actual condition and the desired condition,

4. A means to amplify this difference (which is often small) and use it to move the actual condition towards the desired condition.

In the example of the car, (1) are eyes, (2) is the optic nerve and pathways to the brain,(3) is the brain, and (4) are the arms and steering wheel. A small turn of the wheel translates into a major turn for the car.

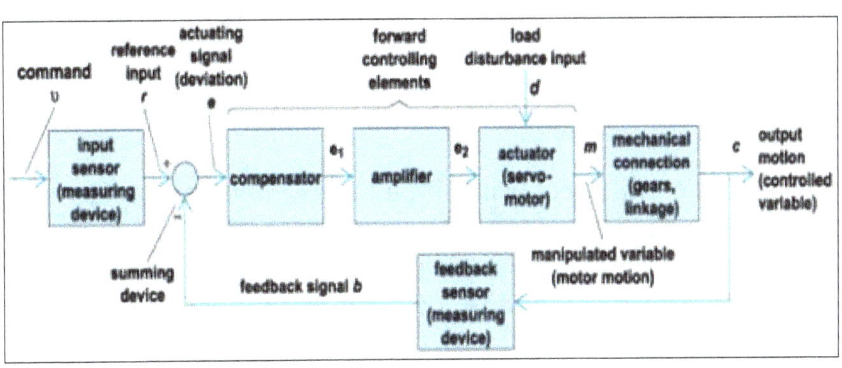

Servo loop elements and their interconnections. Cause-and-effect action takes place in the directions of arrows. (After American National Standards Institute, Terminology for Automatic Control, ANSI C85.1).

The illustration shows the basic elements of a servomechanism and their interconnections; in this type of block diagram the connection between elements is such that only a unidirectional cause-and-effect action takes place in the direction shown by the arrows. The arrows form a closed path or loop; hence this is a single-loop servomechanism or, simply, a servo loop. More complex servomechanisms may have two or more loops (multiloop servo), and a complete control system may contain many servomechanisms.

All servomechanisms have at least these basic components: a controlled device, a command device, an error detector, an error-signal amplifier, and a device to perform any necessary error corrections (the servomotor). In the controlled device, that which is being regulated is usually position. This device must, therefore, have some means of generating a signal (such as a voltage), called the feedback signal, that represents its current position. This signal is sent to an error-detecting device. The command device receives information, usually from outside the system, that represents the desired position of the controlled device. This information is converted to a form usable by the system (such as a voltage) and is fed to the same error detector as is the signal from the controlled device. The error detector compares the feedback signal (representing actual position) with the command signal (representing desired position). Any discrepancy results in an error signal that represents the correction necessary to bring the controlled device to its desired position. The error-correction signal is sent to an amplifier, and the amplified voltage is used to drive the servomotor, which repositions the controlled device. (Servomechanism may or may not use a servomotor. For example a household furnace controlled by thermostat is a servomechanism, yet there is no motor being controlled directly by the servomechanism).

In many applications, servomechanisms allow high-powered devices to be controlled by signals from devices of much lower power. The operation of the high-powered device results from a signal (called the error, or difference, signal) generated from a comparison of the desired position of the high-powered device with its actual position. The ratio between the power of the control signal and that of the device controlled can be on the order of billions to one.

Applications

Servomechanisms are useful to control motion without human attendants, or to maintain the accuracy of an environment like a power plant, and to control action from a remote isolated station. The controller typically uses (and has) much less power than that of what is being controlled. Almost always it is the position or velocity which is being controlled.

Servomechanisms are used to control mechanical things such as motors, steering mechanisms, and robots. Servomechanisms are used extensively in robotics. A robot controller can tell a servomechanism to move in certain ways that depend on the inputs from sensors. Multiple servomechanisms, when interconnected and controlled by a sophisticated computer, can do complex tasks such as cook a meal. A set of servomechanisms, including associated circuits and hardware, and intended for a specific task, constitutes a servo system. Servo systems do precise, often repetitive, mechanical chores. A computer can control a servo system made up of many servomechanisms. For example, an unmanned robotic warplane (also known as a drone) can be programmed to take off, fly a mission, return, and land. Servo systems can be programmed to do assembly-line work and other tasks that involve repetitive movement, precision, and endurance.

A servo robot is a robot whose movement is programmed into a computer. The robot follows the instructions given by the program, and carries out precise motions on that basis. Servo robots can be categorized according to the way they move. In continuous-path motion, the robot mechanism can stop anywhere along its path. In point-to-point motion, it can stop only at specific points in its path. Servo robots can be easily programmed and reprogrammed. This might be done by exchanging diskettes, by manual data entry, or by more exotic methods such as a teach box. When a robot arm must perform repetitive, precise, complex motions, the movements can be entered into the robot controller's memory. Then, when the memory is accessed, the robot arm goes through all the appropriate movements. A teach box is a device that detects and memorizes motions or processes for later recall.

Stepper Motor

A stepper motor is just another digital device, more precisely a digital DC motor. Stepper or Stepper Motor allows you to select a certain degree of movement. Rather than making a whole spin it can divide the spin into smaller parts.

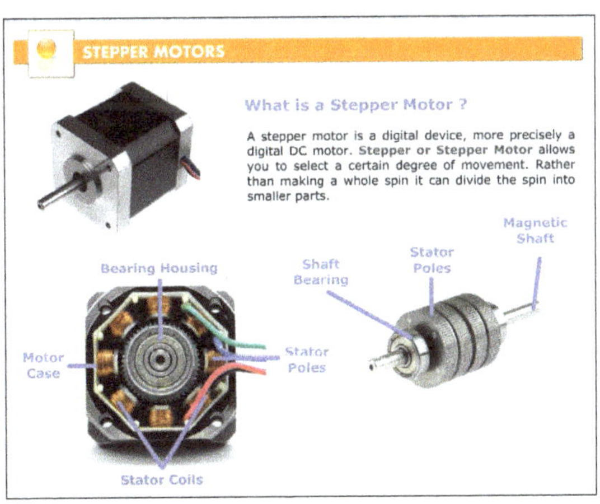

The Stepper Motor can be commanded to hold a certain position for as a long as you want or you can just put a program to control the movements. The Stepper motor is best used in robots where only a certain degree of movement is required. One main problem with ordinary DC Motor is that these keep on moving round and round for infinity amount of time unless you stop feeding them with electricity.

If you want to build a robot that can fetch coffee for you then you can put an ordinary motor in it for moving around but for spreading his arms, collecting the coffee and delivering it safely can only be done by using a stepper motor. But if you put an ordinary motor for controlling its arms then your coffee is probably going to end up on the floor rather than on your table.

Stepper Motor Basics

Every motor converts electricity into motion, the stepper motor or step motor is named so because it converts the electricity into discrete step motion. These steps help you to choose a particular type of motion to perform.

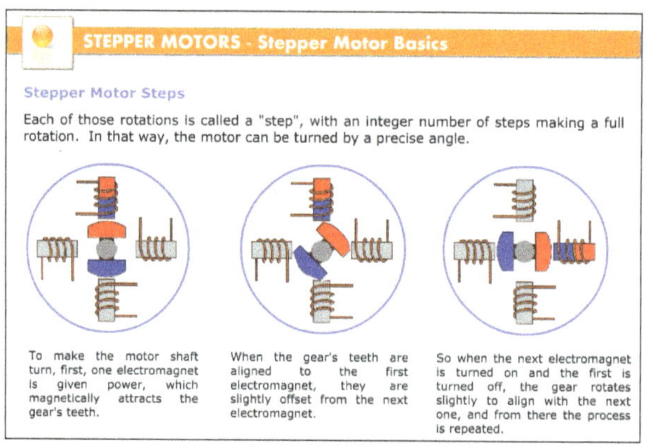

Stepper Motor Steps

DC Motors rotate continuously when voltage is applied to their terminals. Stepper motors, on the other hand, effectively have multiple "toothed" electromagnets arranged around a central gear-shaped piece of iron. The electromagnets are energized by an external control circuit, such as a microcontroller.

The Stepper Motor itself is controlled by a driver which sends the electric pulses to the motor that shows movement in response to these electric pulses. The pulse frequency decides the total amount of movement and rotation. The frequency is controlled by the driver which in turn is controlled by a human that means you are controlling the movement directly.

To make the motor shaft turn, first, one electromagnet is given power, which magnetically attracts the gear's teeth. When the gear's teeth are aligned to the first electromagnet, they are slightly offset from the next electromagnet. So when the next electromagnet is turned on and the first is turned off, the gear rotates slightly to align with the next one, and from there the process is repeated. Each of those rotations is called a "step", with an integer number of steps making a full rotation. In that way, the motor can be turned by a precise angle.

Stepper motors can respond and accelerate quickly. They have low rotor inertia that can get up to speed quickly. For this reason step motors are ideal for short, quick moves.

One Step of a Stepper Motor

A stepper motor (or step motor) is a brushless DC electric motor that divides a full rotation into a number of equal steps. Every turn of the motor is divided into a discrete number of steps, in many cases 200 steps. The stepper motor driver sends a special pulse to the Stepper Motor for each step. Since each pulse causes the motor to rotate a precise angle, typically 1.8°, the motor's position can be controlled without any feedback mechanism. As the digital pulses increase in frequency, the step movement changes into continuous rotation, with the speed of rotation directly proportional to the frequency of the pulses.

Stepper Motor Advantages

- It is very safe to use.
- It is very easy to setup.
- It provides with highly controlled movement.
- The Stepper motors have very long life, only if you don't break it by an accident.
- It is an excellent repeater device it can repeat its movements very accurately.
- Stepper motor is also very cheap as compared to other motion controlling devices.
- One of the main features of stepper motor is that it can't be damaged by overloading.
- When overloaded it stops working until the load is removed.
- It is easy to operate.
- It offers very precise movement.

Stepper Motor Disadvantages

- Resonance occurs if not properly controlled.
- Have low efficiency; it draws more power than the out-put it provides.

- Its torque is inversely proportional to its speed.
- It can't accelerate load rapidly.
- It becomes very difficult to operate at high speeds.

Stepper Motor Accuracy

One of the most remarkable features of the step motor is its precision, the stepper motors are very accurate. But like all other human-made things these have some errors. The precision is not totally awe-inspiring it has chances of error. Standard step motors have an accuracy of ± 3 arc minutes (0.05°).

The remarkable feature of steps motors, though, is that this error does not accumulate from step to step. When a standard step motor travels one step, it will go 1.8° ± 0.05°. If the same motor travels one million steps, it will travel 1,800,000° ± 0.05°. The error does not accumulate. Mostly the error percentage is about pulse minus 0.005. These error chances remain same whether motor takes ten steps or million.

Stepper Motor Applications

Step motors are used every day in both industrial and commercial applications because of their low cost, high reliability, high torque at low speeds. For making more advanced technologies we use stepper motors rather than simple motors. Computer-controlled stepper motors are a type of motion-control positioning system. Step Motors are typically digitally controlled motors, for use in holding or precise positioning applications. These motors are used at:

-Industrial Machines: CNC Machines, Milling Machines, Laser Cutters etc.

-Computer Technology: CD-Roms, DVD Players, Floppy Disk Drives, Scanners etc.

-Printing: Printers, Plotters, 3D Printers etc.

-Intelligent Lighting Systems: Lasers, Optical Devices, Mirror Mounts.

Locomotion in Robots

Locomotion involves the conversion of some source of energy — electricity, air pressure, steam, nuclear power. — into a mechanical action that moves a vehicle or other carriage. Consider the lowly car: gas you put into the tank is converted to mechanical power by means of internal combustion. The gas is compressed as a vapor and explodes against a cylinder. The explosion pushes the cylinder down; this cylinder is in turn is connected to a drive shaft, which spins the wheels. The process repeats itself thousands of times per minute.

Mobile robots use a variety of techniques to achieve motion. Most use an electric power source (usually a battery) that operates an electric motor. In the typical arrangement the direction of the motors can be changed, allowing the robot to be propelled forward or backward. There are other power train techniques used for robots, but the battery and motor pair is by far the most common.

Wheels

Wheels are the most popular method of providing robot mobility. Robot wheels can be just about any size, dictated only by the dimensions of the robot and your outlandish imagination. However, for reasons of practicality and weight, small robots usually have small wheels, less than two or three inches in diameter. Medium sized robots use wheels with diameters up to seven or eight inches. A few unusual designs call for bicycle wheels, which despite their size are lightweight but very sturdy.

Robots can have just about any number of wheels, although two is the most common. The robot is balanced on its wheels by one or two skids or swivel casters. Four and six wheel robots are also around. The most common robot uses two drive wheels. Placed on either side.

Legs

A small percentage of amateur robots are designed with legs, and such robots can be conversation pieces all of their own. Many difficulties must be overcome in designing and constructing a legged robot. First, there is the question of the number of legs, and how the legs provide stability when the robot is in motion. Or more critical, stability when it's standing still. Then there is the question of how the legs propel the robot forward or backward – and more difficult still! – The question of how to turn the robot so it can navigate a corner.

Two-legged walking robot.

Legged robots are a challenge to design and build, but they provide an extra level of mobility that wheeled robots cannot. Wheel-based robots may have a difficult time navigating through rough terrain, but a properly designed leg-based robot can easily walk right over small ditches and obstacles.

A few daring robot experimenters have come out with two-legged robots, but the difficulties in assuring balance and control make these designs largely impractical for most robot hobbyists. Exception: the robot uses wide “duck” feet, and waddles as it walks. With this arrangement, however, turning can be inexact.

Four-legged robots (quadrupeds) are easier to balance, but good locomotion and steering can be difficult to achieve. I’ve found robots with six legs (called hexapods) are able to walk at brisk speeds without falling, and are more than capable of turning corners, bounding over uneven terrain, and making the neighborhood dogs and cats hide for cover.

Robots with legs add many complexities.

Tracks

The basic design of track driven robots is simple: two tracks, one on each side of the robot, act as giant wheels. The tracks turn, and the robot lurches forward or backward. For maximum traction, each track is the same length, or somewhat shorter, than the length of the robot itself — though many variations are possible.

Tankbot Servo from Budget Robotics.

Track drive is practical for many reasons, including the ability to mow through all sorts of obstacles, like rocks, ditches, and potholes. Given the right track material, traction is excellent, even on slippery surfaces like snow, wet concrete, or a clean kitchen floor.

For the most part, constructing and effective track drive is harder than implementing wheels. The reason: the tracks present a large contact area. This larger contact area increases traction when moving forward or backward, but it also restricts turning. Tracked vehicles, like tanks, turn by skidding or slipping around a turning point – hence they are referred to as having skid-steering. If the treads are super-pliable, and the surface is hard (like a kitchen floor), the added friction can greatly impair the ability of the vehicle to turn.

Zero Moment Point

Zmp for Bipedal Robots

The Zero-Moment Point, or ZMP, is the point on the surface of the foot where a resultant force R can replace the force distribution shown in figure below. Mathematically, the ZMP can be calculated from a group of contact points p_i for $i = 1, \ldots, N$ with each force vector f_i associated with the contact point,

$$ZMP = \frac{\sum_{i=1}^{N} p_i f_{iz}}{\sum_{i=1}^{N} f_{iz}} = \frac{\sum_{i=1}^{N} p_i f_{iz}}{f_z}.$$

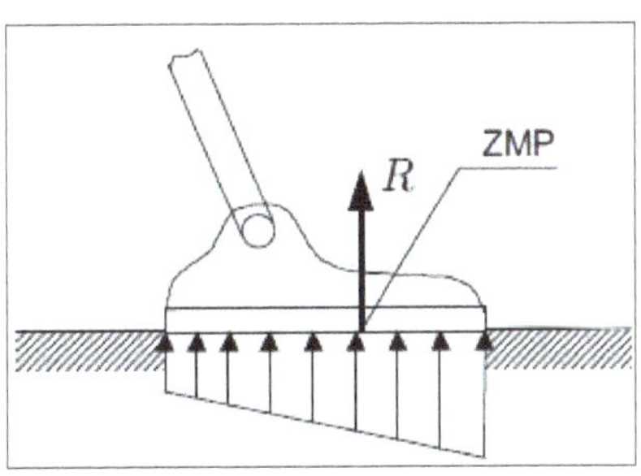

Figure. Definition of Zero-Moment Point.

According to this, the ZMP can never leave the support polygon. If the floor is assumed horizontal, the torque reduces to:

$$\tau_x = \tau_y = 0$$

at the ZMP. Further derivations for the torque at the ZMP can be found in Appendix A.1. This definition is useful when pressure sensors are attached to the feet. With these sensors, the center of pressure can be calculated on the feet, and the ZMP can be directly measured. The humanoid robot DARwIn-OP has pressure sensor foot attachments that can be bought separately. However, these sensors were not used, so a model-based method was employed to calculate the ZMP.

Computed ZMP

Instead of calculating the resultant force from the force distribution, the position of the ZMP can be calculated from the dynamics of the system using Newton's Second Law when the mass properties and motion of the robot are known. To illustrate the use of computed ZMP, the cart-table model used in the controller is presented in figure. This model includes a cart of mass M that rolls without slipping along a flat table with a height of z_c. The table does not rotate as long as point p, denoting the ZMP, stays inside the base, inferring that $\tau_{zmp} = 0$. If the horizontal position of the center of mass is given by x relative to origin O, the net torque around point p can be calculated from the vector equation:

$$\tau_p = p \times R + \tau_{zmp}$$

This expression is simplified to the scalar equation:

$$\tau_{zmp} = -Mg\left(x - p\right) + Mz_c\ddot{x}$$

Using Eq. $\tau_x = \tau_y = 0$, Eq. $\tau_{zmp} = -Mg\left(x - p\right) + Mz_c\ddot{x}$ can be simplified and solved to give an expression for the ZMP,

$$p = x - \frac{z_c}{g}\ddot{x}$$

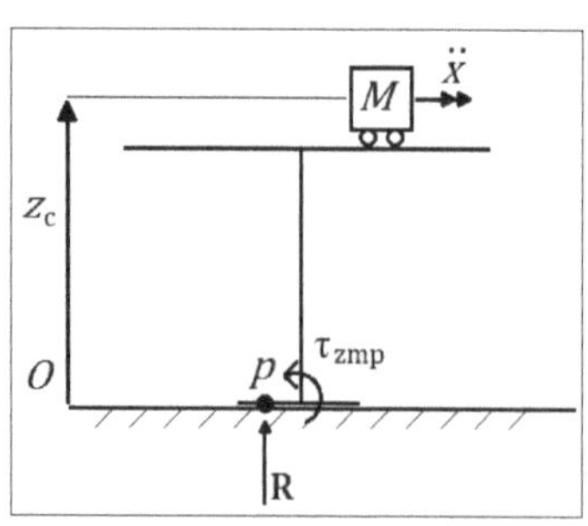

Figure: Cart-Table Model.

Looking closer at Eq. $\tau_{zmp} = -Mg\left(x - p\right) + Mz_c\ddot{x}$, two interesting observations emerge about the computed ZMP from this definition.

1. The ZMP becomes the CoM when there is no acceleration.
2. The ZMP can be located outside the support polygon.

If the ZMP leaves the support polygon during the gait, the motion is determined to be unstable because it can cause the robot to tip over about an edge of the support polygon. This, however, does not necessarily mean that the robot will fall. For example, if a bipedal robot has feet glued to the ground, the ZMP could be outside the support polygon without the robot falling. The cohesive force of the glue adds an extra force to balance the torque around the ZMP so that the robot does not fall until the glue fails. If falling, the robot can recover stable walking either through its motion (compensating body movement) or increasing its support polygon (placing the other foot on the ground).

The ZMP can also be calculated for the 3D case using Newton's and Euler's laws for the change in linear and angular momentum. More general expressions, similar to Eq. 1.5, were presented in and summarized in Appendix A.2. This was used in modeling the full 3D dynamics of DARwIn-OP to calculate the ZMP. However, this 3D model was never formally used in the controller, but was developed using Kane's method.

The Human Gait and a Simplified Model

One of the goals of implementing this humanoid walking controller is to achieve a walking gait that is more human-like. In order to achieve this, the human gait is presented and then simplified to match the needs of the controller. During walking, at least one foot remains in contact with the ground at all times. Each step consists of two main phases, stance and swing, consisting of about 62% and 38% of the gait cycle, respectively. Each of these phases can be subdivided into 8 more categories : initial contact, loading response, midstance, terminal stance, preswing, initial swing, midswing, and terminal swing as shown in figure.

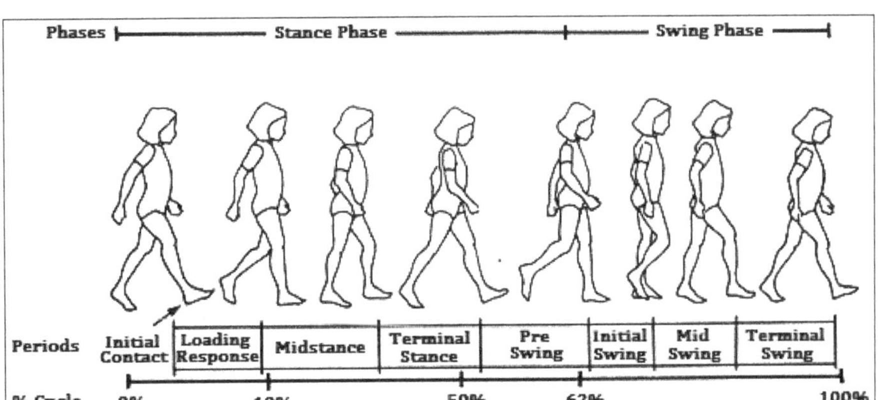

Figure: Human Gait Cycle.

When the leading foot touches the ground, initial contact begins, starting the stance phase. Usually, this involves a simple heel strike to the ground. The main implusive force from ground contact is experienced in the next phase, the loading response. In this phase, the leading foot flattens, coming in full contact with the ground. The leading foot then absorbs the contact force from the ground. After this, the first single support phase, called midstance, begins when the trailing foot leaves the ground and ends when the body weight is aligned over the front of the current stance foot. During midstance, the body has a maximum potential energy as the body's center of gravity reaches its maximum height, rising over the stance leg. The terminal stance phase begins as soon as the body's weight shifts past the current stance foot. Heel off occurs during this phase when the stance foot's heel leaves the ground. The final stance phase is preswing, in which the previous stance foot prepares to lift off, but is still in contact with the ground. This is called toe off as the previous stance foot's toe leaves the ground and can no longer provide active forward propulsion.

Now, the previous stance leg during single support becomes the swing foot for the next three phases. Initial swing begins immediately after the swing foot leaves the ground and lasts until maximum knee flexion, where midswing starts. During midswing, the swing foot is brought past the stance foot, making sure that the foot has enough ground clearance. The phase ends when the tibia is perpendicular to the ground. The final phase, terminal swing, prepares the body for contact with the ground and fully extends the knee. This completes one gait cycle.

In order to simplify the human gait cycle, the proposed controller is broken into four distinct phases: right leg double support, right leg single support, left leg double support, and left leg single support. Relating these to the aforementioned phases, right leg double support tries to capture initial contact to midstance. Right leg single support refers to the period of time during which the left leg is swinging through the air, namely midstance and most of terminal stance. As soon as the left swing leg touches the ground, left leg double support begins, which includes the last part of terminal stance and preswing. The final part of the gait, left leg single support, includes the entire swing phase from initial swing to terminal swing. The controller develops schemes to handle these four phases throughout the walking gait cycle.

Robotic Non-destructive Testing

Non-destructive testing (NDT) is a highly multidisciplinary group of analysis techniques used throughout science and industry to evaluate the properties of materials, and/or to ensure the integrity of components/structures, without causing damage to them. In civil aerospace manufacturing, the increasing deployment of composite materials demands a high integrity and traceability of NDT measurements, combined with a rapid throughput of data. Modern components increasingly present challenging shapes and geometries for inspection. Using traditional manual inspection approaches produce a time-consuming bottleneck in the industrial production and this limitation provides the fundamental motivation for increased automation.

Modern Computer-Aided Design (CAD) is used extensively in composite manufacture. Additionally, where it was once necessary to construct large items from many smaller parts, Computer-Aided Manufacturing (CAM) now allows these large items to be produced easily from one piece of raw material (through traditional subtractive approaches, or built up using more recent additive manufacturing processes). As a result, large components with complex geometries are becoming very common in modern structures, and the aerospace industry is a typical field, where wide complex shaped parts are very frequently used. Moreover the use of composite materials, which are notoriously challenging to inspect, is becoming widespread in the construction of new generations of civilian aircraft. To cope with future demand projections for these operations, it is therefore essential to overcome the current NDT bottleneck, which traditionally can be the slowest aspect in a production process.

A fundamental issue with composites manufacturing compared to conventional light alloy materials lies in the process variability. Often parts that are designed as identical, will have significant deviations from CAD, and also may change shape when removed from the mould. This presents a significant challenge for precision NDT measurement deployment which must be flexible to accommodate these manufacturing issues.

For these reasons, NDT inspection is often performed manually by technicians who typically have to position and move appropriate probes over the contour of the sample surfaces. Manual scanning requires trained technicians and results in a very slow inspection process for large samples. The repeatability of a test can be challenging in structures where complex setups are necessary to perform the inspection (e.g. orientation of the probe, constant standoff, etc.). While manual scanning may remain a valid approach around the edges of a structure, or the edges of holes in a structure, developing reliable automated solutions has become an industry priority to drive down inspection times. The fundamental aims of automation within the inspection process are to minimize downtimes due to the higher achievable speed, and to minimize variability due to human factors.

Semi-automated inspection systems have been developed to overcome some of the shortcomings with manual inspection techniques, using both mobile and fixed robotic platforms. The use of linear manipulators and bridge designs has, for a number of years, provided the most stable conditions in terms of positioning accuracy. The use of these systems to inspect parts with noncomplex shapes (plates, cylinders or cones) is widespread; typically, they are specific machines, which are used to inspect identically shaped and/or sized parts.

More recently, many manufacturers of industrial robots have produced robotic manipulators with excellent positional accuracy and repeatability. An industrial robot is defined as an automatically controlled, reprogrammable, multipurpose manipulator, programmable in three or more axes . In the spectrum of robot manipulators, some modern robots have suitable attributes to develop automated NDT systems and cope with the challenging situations seen in the aerospace industry. They present precise mechanical systems, the possibility to accurately master each joint, and the ability to export positional data at frequencies up to 500 Hz. Some applications of 6-axis robotic arms in the NDT field have been published during the last few years and there is a growing interest in using such automation solutions with many manufacturers within the aerospace sector.

Exploring the current state of the art, RABIT is a group of systems developed by TECNATOM S.A., in collaboration with KUKA Robots Ibérica, that first approached the possibility of incorporating the use of industrial robots in NDT applications . These systems boast the capability of using the potential of industrial robots and integrating them in an overall inspection apparatus, bringing together all the hardware and software required to plan and configure ultrasonic inspections. Off-the-shelf robotic arms were also used in the Laser Ultrasound for Composite InspEction (LUCIE) system, addressed to inspect large curved surfaces such as the inside of aircraft fuselage, by means of ultrasound generated by laser. Genesis Systems Group has developed the NSpect family of Robotic Non-Destructive Inspection cells. Incorporating the FlawInspecta technology, developed by Diagnostic Sonar in conjunction with National Instruments, the NSpect systems employ a KUKA 6DOF robot arm to perform ultrasonic inspection using either an immersion tank, or a recirculating water couplant. General Electric (GE) has also investigated the integration of phased array UT with off-the-shelf industrial robots for the inspection of aerospace composites.

Despite these previous efforts, there remain challenges to be addressed before fully automated NDT inspection of complex geometry composite parts becomes commonplace. The key challenges include generation and in-process modification of the robot tool-path, high speed NDT data collection, integration of surface metrology measurements, and overall visualization of measurement results in a user friendly fashion. Collaborations driving this vision include TWI Technology Centre (Wales), which is currently carrying out a 3-year project, called IntACom, on behalf of its sponsors; its objective is to achieve a fourfold increase in the throughput of aerospace components. Additionally the UK RCNDE consortium conducts research into integration of metrology with NDT inspection. Both these consortia have identified the requirement for optimal tool path generation over complex curved surfaces.

Approaches to Robotic Path Planning for NDT Applications

Existing Robotic Path Planning Software

Six-axis robotic arms have traditionally been used in production lines to move the robot end-effector from one position to another for repetitive assembly and welding operations. In this scenario, where the exact trajectory between two points in the space is not too important, the teach pendant of a robot is used to move manually the end-effector to the desired position and orientation at each stage of the robot task. Relevant robot configurations are recorded by the robot controller and a robot programme is then written to command the robot to move through the recorded end-effector postures. More recently, accurate mechanical joints and control units have made industrial robotic arms flexible and precise enough for finishing tasks in manufacturing operations. Robotic manipulators are highly complex systems and the trajectory accuracy of a machining tool has a huge impact on the quality and tolerances of the finished surfaces. As a result, many software environments have been developed by manufacturers, academic researchers and also by the robot manufacturers themselves, in order to help technicians and engineers to programme complex robot tasks. The use of such software platforms to programme robot movements is known as off-line programming (OLP). It is based on the 3D virtual representation of the complete robot work cell, the robot end-effector and the samples to be manipulated or machined. Although some limited applications for inspection delivery have been demonstrated, in general conventional OLP is geared towards manufacturing applications where the task is the production of a specific component using conventional milling/ drilling / trimming operations. In contrast, the result of an automated NDT inspection requires a flexible and extensible approach that has the flexibility to allow future changes in the path planning to accommodate requirements of future NDT inspections. Building the NDT toolbox in MATLAB provides an easy route to such future adaptation by the research community (e.g. fluid dynamics modelling of water-jets, compensation for part variability and conditional programming approaches).

Using current OLP software to generate appropriate tool-paths for NDT purposes can appear quite straightforward at first inspection; however it is possible to list a series of serious inadequacies:

1. Path-planning for automated NDT inspections is a very specific task. As previously mentioned, much commercial software for off-line robot programming draws its origin from the need to use the advantageous flexibility of general robotic manipulators to replace the more traditional and usual machining tools (milling machines, lathes, etc.). As a result, many commercial software applications for off-line programming of robots are expensive tools, incorporating a lot of functionality specific for CAD/CAM purposes and machining features. Conventional OLP software has no easy provision for modification of tool path based on specific characteristics of NDT inspection. For example, in water-jet coupled ultrasonic testing, the exact orientation and separation from part of the end effector, combined with water pressure and nozzle characteristics will influence the trajectory of the water jet. Since the new toolbox is MATLAB based, it is flexible and easy to accommodate such new features into the path planning algorithm in future.

2. Significant complications exist when two or more robotic arms need to be synchronized in order to perform a specific NDT inspection. The Ultrasonic Through-Transmission (UTT) technique, for example, uses two transducers: one emitter and one receiver; the receiver being placed on the opposite side of the component and facing the transmitting probe. Many commercial pieces of software (e.g. Delcam and Mastercam) do not offer any support for co-operating robots. Fast Surf, an add-on from CENIT for Delmia (Dassault Systems), allows partial synchronization of robotic movements (e.g. at start or end points of complex paths) but not full synchronization over the complete path, required for the UTT technique. Our new MATLAB toolbox provides the capability for full point to point synchronization between robots for situations where a change in material section is encountered – this is far more sophisticated than simple master–slave synchronization implemented by typical robot equipment suppliers.

Robotic Sensors

Robotic sensors are used to estimate the robot's condition and environment, These signals are passed to the controller to enable the appropriate behavior, The sensors in the robots are based on the functions of the human sensory organs, The robots require the extensive information about their environment in order to function effectively.

Robots need Robot Sensors to know the world around them, There are many Robot Sensors including ultrasonic, the temperature & the humidity, the force and a lot more to increase the robot awareness.

Types of Robot Sensors

Light Sensors

A Light sensor is used to detect light and create a voltage difference. The two main light sensors generally used in robots are Photoresistor and Photovoltaic cells. Other kinds of light sensors like Phototubes, Phototransistors, CCD's etc. are rarely used.

Photoresistor is a type of resistor whose resistance varies with change in light intensity; more light leads to less resistance and less light leads to more resistance. These inexpensive sensors can be easily implemented in most light dependant robots.

Photovoltaic cells convert solar radiation into electrical energy. This is especially helpful if you are planning to build a solar robot. Although photovoltaic cell is considered as an energy source, an intelligent implementation combined with transistors and capacitors can convert this into a sensor.

Photo Voltaic Cells (Solar Cells) Photoresistors (LDR)

Sound Sensor

As the name suggests, this sensor (generally a microphone) detects sound and returns a voltage proportional to the sound level. Sound Sensor A simple robot can be designed to navigate based on the sound it receives. Imagine a robot which turns right for one clap and turns left for two claps. Complex robots can use the same microphone for speech and voice recognition.

Sound Sensor (Mic)

Implementing sound sensors is not as easy as light sensors because Sound sensors generate a very small voltage difference which should be amplified to generate measurable voltage change.

Temperature Sensors

What if your robot has to work in a desert and transmit ambient temperature? Simple solution is to use a temperature sensor. Tiny temperature sensor ICs provide voltage difference for a change in temperature. Few generally used temperature sensor IC's are LM34, LM35, TMP35, TMP36, and TMP37.

Temperature Sensors

Contact Sensor

Contact sensors are those which require physical contact against other objects to trigger. A push button switch, limit switch or tactile bumper switch are all examples of contact sensors. Limit Switch These sensors are mostly used for obstacle avoidance robots. When these switches hit an obstacle, it triggers the robot to do a task, which can be reversing, turning, switching on a LED, Stopping etc. There are also capacitive contact sensors which react only to human touch (Not sure if they react to animals touch). Touch screen Smart phones available these days use capacitive touch sensors. Contact Sensors can be easily implemented, but the drawback is that they require physical contact. In other words, your robot will not turn until it hits an object. A better alternative is to use a proximity sensor.

Limit Switch (Contact Sensor)

Proximity Sensor

This is a type of sensor which can detect the presence of a nearby object within a given distance, without any physical contact. The working principle of a Proximity sensor is simple. A transmitter transmits an electromagnetic radiation or creates an electrostatic field and a receiver receives and analyzes the return signal for interruptions. There are different types of Proximity sensors and we will discuss only a few of them which are generally used in robots.

- Infrared (IR) Transceivers: An IR LED transmits a beam of IR light and if it finds an obstacle, the light is simply reflected back which is captured by an IR receiver. Few IR transceivers can also be used for distance measurement.
- Ultrasonic Sensor: These sensors generate high frequency sound waves; the received echo suggests an object interruption. Ultrasonic Sensors can also be used for distance measurement.
- Photoresistor: Photoresistor is a light sensor; but, it can still be used as a proximity sensor. When an object comes in close proximity to the sensor, the amount of light changes which in turn changes the resistance of the Photoresistor. This change can be detected and processed.

There are many different kinds of proximity sensors and only a few of them are generally preferred for robots. For example, Capacitive Proximity sensors are available which detects change in capacitance around it. Inductive proximity sensor detects objects and distance through the use of induced magnetic field.

Distance Sensor

Most proximity sensors can also be used as distance sensors, or commonly known as Range Sensors; IR transceivers and Ultrasonic Sensors are best suited for distance measurement.

- Ultrasonic Distance Sensors: The sensor emits an ultrasonic pulse and is captured by a receiver. Since the speed of sound is almost constant in air, which is 344m/s, the time between send and receive is calculated to give the distance between your robot and the obstacle. Ultrasonic distance sensors are especially useful for underwater robots.
- Infrared Distance sensor: IR circuits are designed on triangulation principle for distance measurement. A transmitter sends a pulse of IR signals which is detected by the receiver if there is an obstacle and based on the angle the signal is received, distance is calculated. SHARP has a family of IR transceivers which are very useful for distance measurement. A simple transmit and receive using a couple of transmitters and receivers will still do the job of distance measurement, but if you require precision, then prefer the triangulation method.
- Laser range Sensor: Laser light is transmitted and the reflected light is captured and analyzed. Distance is measured by calculating the speed of light and time taken for the light to reflect back to the receiver. These sensors are very useful for longer distances.
- Encoders: These sensors (not actually sensors, but a combination of different components) convert angular position of a shaft or wheel into an analog or digital code. The most popular encoder is an optical encoder which includes a rotational disk, light source and a light detector (generally an IR transmitter and IR receiver). The rotational disk has transparent and opaque pattern (or just black and white pattern) painted or printed over it. When the disk rotates along with the wheel the emitted light is interrupted generating a signal output. The number of times the interruption happens and the diameter of the wheel can together give the distance travelled by the robot.
- Stereo Camera: Two cameras placed adjacent to each other can provide depth information using its stereo vision. Processing the data received from a camera is difficult for a robot with minimal processing power and memory. If opted for, they make a valuable addition to your robot.

There are other stretch and bend sensors which are also capable of measuring distance. But, their range is so limited that they are almost useless for mobile robots.

Pressure Sensors

As the name suggests, pressure sensor measures pressure. Tactile pressure sensors are useful in robotics as they are sensitive to touch, force and pressure. If you design a robot hand and need to measure the amount of grip and pressure required to hold an object, then this is what you would want to use.

Tilt Sensors

Tilt sensors measure tilt of an object. In a typical analog tilt sensor, a small amount of mercury is suspended in a glass bulb. When mercury flows towards one end, it closes a switch which suggests a tilt.

Navigation/Positioning Sensors

The name says it all. Positioning sensors are used to approximate the position of a robot, some for indoor positioning and few others for outdoor positioning.

- GPS (Global Positioning System): The most commonly used positioning sensor is a GPS. Satellites orbiting our earth transmit signals and a receiver on a robot acquires these signals and processes it. The processed information can be used to determine the approximate position and velocity of a robot. These GPS systems are extremely helpful for outdoor robots, but fail indoors. They are also bit expensive at the moment and if their prices fall, very soon you would see most robots with a GPS module attached.
- Digital Magnetic Compass: Similar to a handheld magnetic compass, Digital Magnetic compass provides directional measurements using the earth's magnetic field which guides your robot in the right direction to reach its destination. These sensors are cheap compared to GPS modules, but a compass works best along with a GPS module if you require both positional feedback and navigation. Philips KMZ51 is sensitive enough to detect earth's magnetic field.
- Localization: Localization refers to the task of automatically determining the location of a robot in complex environment. Localization is based on external elements called landmarks which can be either artificially placed landmarks, or natural landmark. In the first approach, artificial landmarks or beacons are placed around the robot, and a robot's sensor captures these signals to determine its exact location. Natural landmarks can be doors, windows, walls, etc. which are sensed by a robots sensor / vision system (Camera). Localization can be achieved using beacons which generate Wi-Fi, Bluetooth, Ultrasound, Infrared, Radio transmissions, Visible Light, or any similar signal.

Acceleration Sensor

An accelerometer is a device which measures acceleration and tilt. There are two kinds of forces which can affect an accelerometer: Static force and Dynamic Force.

Static Force: Static force is the frictional force between any two objects. For example earth's gravitational force is static which pulls an object towards it. Measuring this gravitational force can tell you how much your robot is tilting. This measurement is exceptionally useful in a balancing robot, or to tell you if your robot is driving uphill or on a flat surface.

Dynamic force: Dynamic force is the amount of acceleration required to move an object. Measuring this dynamic force using an accelerometer tells you the velocity/speed at which your robot is moving. We can also measure vibration of a robot using an accelerometer, if in any case you need to.

Accelerometer comes in different flavors. Always select the one which is most appropriate for your robot. Some of the factors which you need to consider before selecting an accelerometer are:

1. Output Type: Analog or Digital.
2. Number of Axis: 1,2 or 3.
3. Accelerometer Swing: ±1.5g, ±2g, ±4g, ±8g, ±16g.
4. Sensitivity: Higher or Lower (Higher the better).
5. Bandwidth.

Gyroscope

A gyroscope or simply Gyro is a device which measures and helps maintain orientation using the principle of angular momentum. In other words, a Gyro is used to measure the rate of rotation around a particular axis. Gyroscope is especially useful when you want your robot to not depend on earth's gravity for maintaining Orientation. (Unlike accelerometer).

IMU

Inertial Measurement Units combine properties of two or more sensors such as Accelerometer, Gyro, Magnetometer, etc., to measure orientation, velocity and gravitational forces. In simple words, IMU's are capable of providing feedback by detecting changes in an objects orientation (pitch, roll and yaw), velocity and gravitational forces. Few IMUs go a step further and combine a GPS device providing positional feedback.

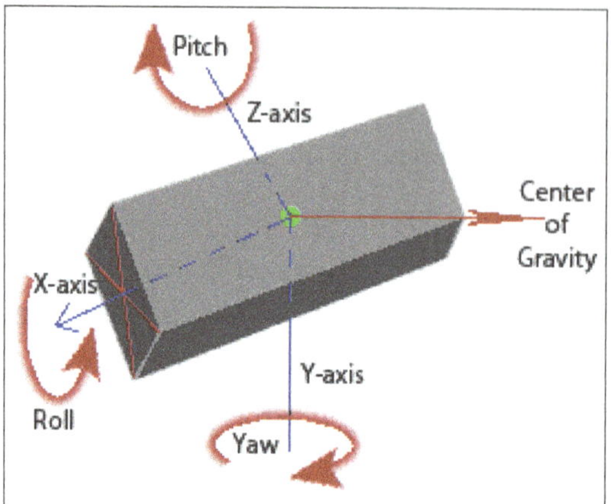

Pitch, Roll and Yaw of an Object

Voltage Sensors

Voltage sensors typically convert lower voltages to higher voltages, or vice versa. One example is a general Operational-Amplifier (Op-Amp) which accepts a low voltage, amplifies it, and generates a higher voltage output. Few voltage sensors are used to find the potential difference between two ends (Voltage Comparator). Even a simple LED can act as a voltage sensor which can detect a voltage difference and light up. (not considering current requirements here).

Current Sensors

Current sensors are electronic circuits which monitor the current flow in a circuit and output either a proportional voltage or a current. Most current sensors output an analog voltage between 0V to 5V which can be processed further using a microcontroller.

Other Sensors for Robots

There are hundreds of sensors made today to sense virtually anything you can think of, and it is almost impossible to list all available sensors. Apart from those mentioned above, there are many other sensors used for specific applications. For example: Humidity Sensors measures Humidity; Gas sensors are designed to detect particular gases (helpful for robots which detects gas leaks); Potentiometers are so versatile that they can be used in numerous different applications; Magnetic Field Sensors detect the strength of magnetic field around it.

References

- Tools-as-end-effectors: roboticsbible.com, Retrieved 17 July 2020
- Stepper-Motors: robotpark.com, Retrieved 14 June 2020

- Types-of-robotic-actuators-8253: roboticstomorrow.com, Retrieved 12 May 2020
- What-do-delta-robots-have-in-common-with-spiders: linearmotiontips.com, Retrieved 28 April 2020
- Types-of-robot-sensors: robotplatform.com, Retrieved 10 April 2020
- Exploring-robot-locomotion-systems: robotoid.com, Retrieved 24 May 2020

5

Automation in Robotics

Automation involves the use of a wide range of technologies and computer software to perform tasks that are usually done by humans. Vehicles that are capable of sensing their environment and moving safely without human intervention are known as autonomous vehicles. This chapter discusses in detail the theories and concepts related to robotics and automation.

Automation

Automation is, the application of machines to tasks once performed by human beings or, increasingly, to tasks that would otherwise be impossible. Although the term mechanization is often used to refer to the simple replacement of human labor by machines, automation generally implies the integration of machines into a self-governing system. Automation has revolutionized those areas in which it has been introduced, and there is scarcely an aspect of modern life that has been unaffected by it.

The term automation was coined in the automobile industry about 1946 to describe the increased use of automatic devices and controls in mechanized production lines. The origin of the word is attributed to D.S. Harder, an engineering manager at the Ford Motor Company at the time. The term is used widely in a manufacturing context, but it is also applied outside manufacturing in connection with a variety of systems in which there is a significant substitution of mechanical, electrical, or computerized action for human effort and intelligence.

In general usage, automation can be defined as a technology concerned with performing a process by means of programmed commands combined with automatic feedback control to ensure proper execution of the instructions. The resulting system is capable of operating without human intervention. The development of this technology has become increasingly dependent on the use of computers and computer-related technologies. Consequently, automated systems have become increasingly sophisticated and complex. Advanced systems represent a level of capability and performance that surpass in many ways the abilities of humans to accomplish the same activities.

Automation technology has matured to a point where a number of other technologies have developed from it and have achieved a recognition and status of their own. Robotics is one of these technologies; it is a specialized branch of automation in which the automated machine possesses certain anthropomorphic, or humanlike, characteristics. The most typical humanlike characteristic of a modern industrial robot is its powered mechanical arm. The robot's arm can be programmed to move through a sequence of motions to perform useful tasks, such as loading and unloading parts at a production machine or making a sequence of spot-welds on the sheet-metal parts of an automobile body during assembly. As these examples suggest, industrial robots are typically used to replace human workers in factory operations.

Principles and Theory of Automation

The developments described above have provided the three basic building blocks of automation: (1) a source of power to perform some action, (2) feedback controls, and (3) machine programming. Almost without exception, an automated system will exhibit all these elements.

Power Source

An automated system is designed to accomplish some useful action, and that action requires power. There are many sources of power available, but the most commonly used power in today's automated systems is electricity. Electrical power is the most versatile, because it can be readily generated from other sources (e.g., fossil fuel, hydroelectric, solar, and nuclear) and it can be readily converted into other types of power (e.g., mechanical, hydraulic, and pneumatic) to perform useful work. In addition, electrical energy can be stored in high-performance, long-life batteries.

The actions performed by automated systems are generally of two types: (1) processing and (2) transfer and positioning. In the first case, energy is applied to accomplish some processing operation on some entity. The process may involve the shaping of metal, the molding of plastic, the switching of electrical signals in a communication system, or the processing of data in a computerized information system. All these actions entail the use of energy to transform the entity (e.g., the metal, plastic, electrical signals, or data) from one state or condition into another more valuable state or condition. The second type of action—transfer and positioning—is most readily seen in automated manufacturing systems designed to perform work on a product. In these cases, the product must generally be moved (transferred) from one location to another during the series of processing steps. At each processing location, accurate positioning of the product is generally required. In automated communications and information systems, the terms transfer and positioning refer to the movement of data (or electrical signals) among various processing units and the delivery of information to output terminals (printers, video display units, etc.) for interpretation and use by humans.

Feedback Controls

Feedback controls are widely used in modern automated systems. A feedback control system consists of five basic components: (1) input, (2) process being controlled, (3) output, (4) sensing elements, and (5) controller and actuating devices. These five components are illustrated. The term closed-loop feedback control is often used to describe this kind of system.

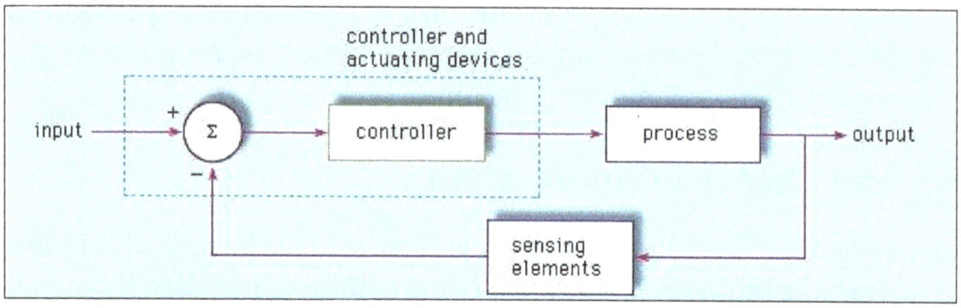

The components of a feedback control system and their relationships.

The input to the system is the reference value, or set point, for the system output. This represents the desired operating value of the output. Using the previous example of the heating system as an illustration, the input is the desired temperature setting for a room. The process being controlled is the heater (e.g., furnace). In other feedback systems, the process might be a manufacturing operation, the rocket engines on a space shuttle, the automobile engine in cruise control, or any of a variety of other processes to which power is applied. The output is the variable of the process that is being measured and compared to the input; in the above example, it is room temperature.

The sensing elements are the measuring devices used in the feedback loop to monitor the value of the output variable. In the heating system example, this function is normally accomplished using a bimetallic strip. This device consists of two metal strips joined along their lengths. The two metals possess different thermal expansion coefficients; thus, when the temperature of the strip is raised, it flexes in direct proportion to the temperature change. As such, the bimetallic strip is capable of measuring temperature. There are many different kinds of sensors used in feedback control systems for automation.

The purpose of the controller and actuating devices in the feedback system is to compare the measured output value with the reference input value and to reduce the difference between them. In general, the controller and actuator of the system are the mechanisms by which changes in the process are accomplished to influence the output variable. These mechanisms are usually designed specifically for the system and consist of devices such as motors, valves, solenoid switches, piston cylinders, gears, power screws, pulley systems, chain drives, and other mechanical and electrical components. The switch connected to the bimetallic strip of the thermostat is the controller and actuating device for the heating system. When the output (room temperature) is below the set point, the switch turns on the heater. When the temperature exceeds the set point, the heat is turned off.

Machine Programming

The programmed instructions determine the set of actions that is to be accomplished automatically by the system. The program specifies what the automated system should do and how its various components must function in order to accomplish the desired result. The content of the program varies considerably from one system to the next. In relatively simple systems, the program consists of a limited number of well-defined actions that are performed continuously and repeatedly in the proper sequence with no deviation from one cycle to the next. In more complex systems, the number of commands could be quite large, and the level of detail in each command could be significantly greater. In relatively sophisticated systems, the program provides for the sequence of actions to be altered in response to variations in raw materials or other operating conditions.

Programming commands are related to feedback control in an automated system in that the program establishes the sequence of values for the inputs (set points) of the various feedback control loops that make up the automated system. A given programming command may specify the set point for the feedback loop, which in turn controls some action that the system is to accomplish. In effect, the purpose of the feedback loop is to verify that the programmed step has been carried out. For example, in a robot controller, the program might specify that the arm is to move to a designated position, and the feedback control system is used to verify that the move has been correctly made. The relationship of program control and feedback control in an automated system is illustrated.

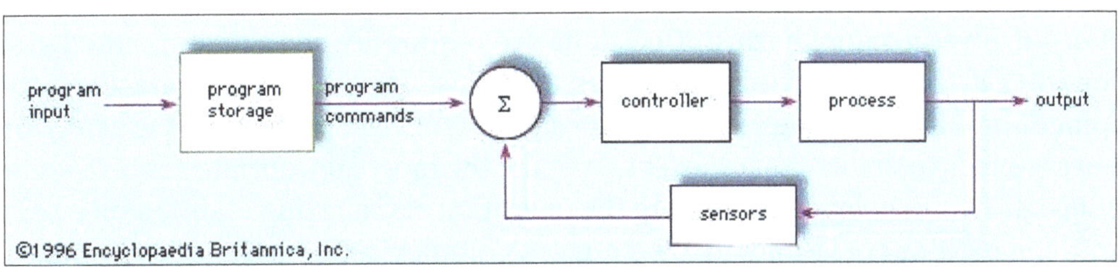

Relationship of program control and feedback control in an automated system.

Some of the programmed commands may be executed in a simple open-loop fashion—i.e., without the need for a feedback loop to verify that the command has been properly carried out. For example, a command to flip an electrical switch may not require feedback. The need for feedback control in an automated system might arise when there are variations in the raw materials being fed into a production process, and the system must take these variations into consideration by making adjustments in its controlled actions. Without feedback, the system would be unable to exert sufficient control over the quality of the process output.

The programmed commands may be contained on mechanical devices (e.g., mechanical cams and linkages), punched paper tape, magnetic tape, magnetic disks, computer memory, or any of a variety of other media that have been developed over the years for particular applications. It is common today for automated equipment to use computer storage technology as the means for storing the programmed commands and converting them into controlled actions. One of the advantages of computer storage is that the program can be readily changed or improved. Altering a program that is contained on mechanical cams involves considerable work.

Programmable machines are often capable of making decisions during their operation. The decision-making capacity is contained in the control program in the form of logical instructions that govern the operation of such a system under varying circumstances. Under one set of circumstances, the system responds one way; under different circumstances, it responds in another way. There are several reasons for providing an automated system with decision-making capability, including (1) error detection and recovery, (2) safety monitoring, (3) interaction with humans, and (4) process optimization.

Error detection and recovery is concerned with decisions that must be made by the system in response to undesirable operating conditions. In the operation of any automated system, malfunctions and errors sometimes occur during the normal cycle of operations, for which some form of corrective action must be taken to restore the system. The usual response to a system malfunction has been to call for human assistance. There is a growing trend in automation and robotics to enable the system itself to sense these malfunctions and to correct for them in some manner without human intervention. This sensing and correction is referred to as error detection and recovery, and it requires that a decision-making capability be programmed into the system.

Safety monitoring is a special case of error detection and recovery in which the malfunction involves a safety hazard. Decisions are required when the automated system sensors detect that a safety condition has developed that would be hazardous to the equipment or humans in the vicinity of the equipment. The purpose of the safety-monitoring system is to detect the hazard and to take the most appropriate action to remove or reduce it. This may involve stopping the operation and alerting maintenance personnel to the condition, or it may involve a more complex set of actions to eliminate the safety problem.

Automated systems are usually required to interact with humans in some way. An automatic bank teller machine, for example, must receive instructions from customers and act accordingly. In some automated systems, a variety of different instructions from humans is possible, and the decision-making capability of the system must be quite sophisticated in order to deal with the array of possibilities.

A fourth reason for decision making in an automated system is to optimize the process. The need for optimization occurs most commonly in processes in which there is an economic performance criterion whose optimization is desirable. For example, minimizing cost is usually an important objective in manufacturing. The automated system might use adaptive control to receive appropriate sensor signals and other inputs and make decisions to drive the process toward the optimal state.

Example of Automation

Development of Robotics

Robotics is based on two related technologies: numerical control and teleoperators. Numerical control (NC) is a method of controlling machine tool axes by means of numbers that have been coded on punched paper tape or other media. It was developed during the late 1940s and early 1950s. The first numerical control machine tool was demonstrated in 1952 in the United States at the Massachusetts Institute of Technology (MIT). Subsequent research at MIT led to the development of the APT (Automatically Programmed Tools) language for programming machine tools.

A teleoperator is a mechanical manipulator that is controlled by a human from a remote location. Initial work on the design of teleoperators can be traced to the handling of radioactive materials in the early 1940s. In a typical implementation, a human moves a mechanical arm and hand at one location, and these motions are duplicated by the manipulator at another location.

Industrial robotics can be considered a combination of numerical-control and teleoperator technologies. Numerical control provides the concept of a programmable industrial machine, and teleoperator technology contributes the notion of a mechanical arm to perform useful work. The first industrial robot was installed in 1961 to unload parts from a die-casting operation. Its development was due largely to the efforts of the Americans George C. Devol, an inventor, and Joseph F. Engelberger, a businessman. Devol originated the design for a programmable manipulator, the U.S. patent for which was issued in 1961. Engelberger teamed with Devol to promote the use of robots in industry and to establish the first corporation in robotics—Unimation, Inc.

The Robot Manipulator

An industrial robot is a reprogrammable, multifunctional manipulator designed to move materials, parts, tools, or specialized devices through variable programmed motions for the performance of a variety of tasks.

The technology of robotics is concerned with the design of the mechanical manipulator and the computer systems used to control it. It is also concerned with the industrial applications of robots, which are described below.

The mechanical manipulator of an industrial robot is made up of a sequence of link and joint combinations. The links are the rigid members connecting the joints. The joints (also called axes) are the movable components of the robot that cause relative motion between adjacent links. As shown in, there are five principal types of mechanical joints used to construct the manipulator. Two of the joints are linear, in which the relative motion between adjacent links is translational, and three are rotary types, in which the relative motion involves rotation between links.

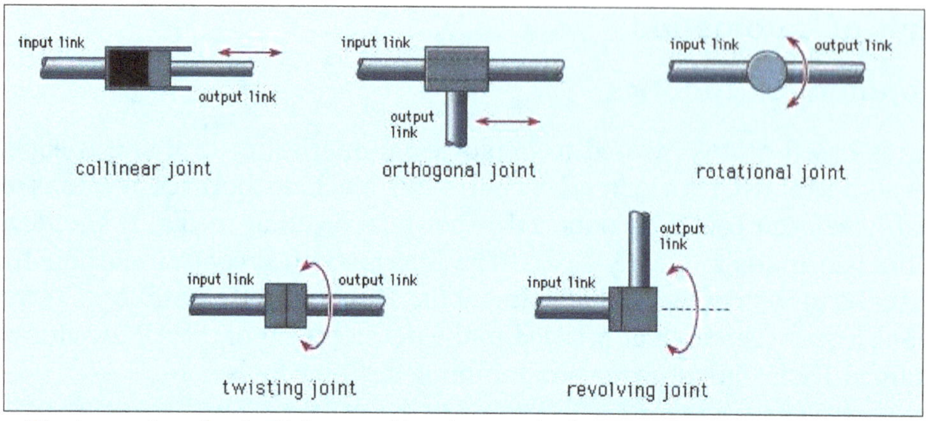

Five types of mechanical joints used in robot manipulators. Collinear and orthogonal are translational joints; rotational, twisting, and revolving are rotary joints.

The manipulator can be divided into two sections: (1) an arm-and-body, which usually consists of three joints connected by large links, and (2) a wrist, consisting of two or three compact joints. Attached to the wrist is a gripper to grasp a work part or a tool (e.g., a spot-welding gun) to perform a process. The two manipulator sections have different functions: the arm-and-body is used to move and position parts or tools in the robot's work space, while the wrist is used to orient the parts or tools at the work location. The arm-and-body section of most commercial robots is based on one of four configurations. Each of the anatomies, as they are sometimes called, provides a different work envelope (i.e., the space that can be reached by the robot's arm) and is suited to different types of applications.

Robot Programming

The computer system that controls the manipulator must be programmed to teach the robot the particular motion sequence and other actions that must be performed in order to accomplish its task. There are several ways that industrial robots are programmed. One method is called lead-through programming. This requires that the manipulator be driven through the various motions needed to perform a given task, recording the motions into the robot's computer memory. This can be done either by physically moving the manipulator through the motion sequence or by using a control box to drive the manipulator through the sequence.

A second method of programming involves the use of a programming language very much like a computer programming language. However, in addition to many of the capabilities of a computer programming language (i.e., data processing, computations, communicating with other computer devices, and decision making), the robot language also includes statements specifically designed for robot control. These capabilities include (1) motion control and (2) input/output. Motion-control commands are used to direct the robot to move its manipulator to some defined position in space. For example, the statement "move P1" might be used to direct the robot to a point in space called P1. Input/output commands are employed to control the receipt of signals from sensors and other devices in the work cell and to initiate control signals to other pieces of equipment in the cell. For instance, the statement "signal 3, on" might be used to turn on a motor in the cell, where the motor is connected to output line 3 in the robot's controller.

Advantages and Disadvantages of Automation

Advantages commonly attributed to automation include higher production rates and increased productivity, more efficient use of materials, better product quality, improved safety, shorter workweeks for labor, and reduced factory lead times. Higher output and increased productivity have been two of the biggest reasons in justifying the use of automation. Despite the claims of high quality from good workmanship by humans, automated systems typically perform the manufacturing process with less variability than human workers, resulting in greater control and consistency of product quality. Also, increased process control makes more efficient use of materials, resulting in less scrap.

Worker safety is an important reason for automating an industrial operation. Automated systems often remove workers from the workplace, thus safeguarding them against the hazards of the factory environment. In the United States the Occupational Safety and Health Act of 1970 (OSHA) was enacted with the national objective of making work safer and protecting the physical well-being of the worker. OSHA has had the effect of promoting the use of automation and robotics in the factory.

Another benefit of automation is the reduction in the number of hours worked on average per week by factory workers. About 1900 the average workweek was approximately 70 hours. This has gradually been reduced to a standard workweek in the United States of about 40 hours. Mechanization and automation have played a significant role in this reduction. Finally, the time required to process a typical production order through the factory is generally reduced with automation.

A main disadvantage often associated with automation, worker displacement, has been discussed above. Despite the social benefits that might result from retraining displaced workers for other jobs, in almost all cases the worker whose job has been taken over by a machine undergoes a period of emotional stress. In addition to displacement from work, the worker may be displaced geographically. In order to find other work, an individual may have to relocate, which is another source of stress.

Other disadvantages of automated equipment include the high capital expenditure required to invest in automation (an automated system can cost millions of dollars to design, fabricate, and install), a higher level of maintenance needed than with a manually operated machine, and a generally lower degree of flexibility in terms of the possible products as compared with a manual system (even flexible automation is less flexible than humans, the most versatile machines of all).

Also there are potential risks that automation technology will ultimately subjugate rather than serve humankind. The risks include the possibility that workers will become slaves to automated machines, that the privacy of humans will be invaded by vast computer data networks, that human error in the management of technology will somehow endanger civilization, and that society will become dependent on automation for its economic well-being.

These dangers aside, automation technology, if used wisely and effectively, can yield substantial opportunities for the future. There is an opportunity to relieve humans from repetitive, hazardous, and unpleasant labour in all forms. And there is an opportunity for future automation technologies to provide a growing social and economic environment in which humans can enjoy a higher standard of living and a better way of life.

Autonomous Car

A self-driving car (sometimes called an autonomous car or driverless car) is a vehicle that uses a combination of sensors, cameras, radar and artificial intelligence (AI) to travel between destinations without a human operator. To qualify as fully autonomous, a vehicle must be able to navigate without human intervention to a predetermined destination over roads that have not been adapted for its use.

Companies developing and/or testing autonomous cars include Audi, BMW, Ford, Google, General Motors, Tesla, Volkswagen and Volvo. Google's test involved a fleet of self-driving cars -- including Toyota Prii and an Audi TT -- navigating over 140,000 miles of California streets and highways.

Levels of Autonomy in Self-driving Cars

The U.S. National Highway Traffic Safety Administration (NHTSA) lays out six levels of automation, beginning with zero, where humans do the driving, through driver assistance technologies up to fully autonomous cars. Here are the five levels that follow zero automation:

- Level 1: Advanced driver assistance system (ADAS) aid the human driver with either steering, braking or accelerating, though not simultaneously. ADAS includes rearview cameras and features like a vibrating seat warning to alert drivers when they drift out of the traveling lane.
- Level 2: An ADAS that can steer and either brake or accelerate simultaneously while the driver remains fully aware behind the wheel and continues to act as the driver.
- Level 3: An automated driving system (ADS) can perform all driving tasks

under certain circumstances, such as parking the car. In these circumstances, the human driver must be ready to re-take control and is still required to be the main driver of the vehicle.

- Level 4: An ADS is able to perform all driving tasks and monitor the driving environment in certain circumstances. In those circumstances, the ADS is reliable enough that the human driver needn't pay attention.
- Level 5: The vehicle's ADS acts as a virtual chauffeur and does all the driving in all circumstances. The human occupants are passengers and are never expected to drive the vehicle.

As of 2018, car makers have reached Level 3. Self-driving vehicles are at least a few years off, because manufacturers must clear a variety of technological milestones and a number of important issues must be addressed before autonomous vehicles can be purchased and used on public roads in the United States, according to the NHTSA.

How Self-driving Cars Work

AI technologies power self-driving car systems. Developers of self-driving cars use vast amounts of data from image recognition systems, along with machine learning and neural networks, to build systems that can drive autonomously.

The neural networks identify patterns in the data, which is fed to the machine learning algorithms. That data includes images from cameras on self-driving cars from which the neural network learns to identify as traffic lights, trees, curbs, pedestrians, street signs and other parts of any given driving environment.

For example, Google's self-driving car project, called Waymo, uses a mix of sensors, Lidar (light detection and ranging -- a technology similar to radar), and cameras and combines all of the data those systems generate to identify everything around the vehicle and predict what those objects might do next. This happens in fractions of a second. The system learns more as it drives, so maturity is important with these systems.

How Google Waymo vehicles work:

- The driver (or passenger) sets a destination. The car's software calculates a route.
- A rotating, roof-mounted Lidar sensor monitors a 60-meter range around the car and creates a dynamic 3D map of the car's current environment.
- A sensor on the left rear wheel monitors sideways movement to detect the car's position relative to the 3D map.
- Radar systems in the front and rear bumpers calculate distances to obstacles.

- AI software in the car is connected to all the sensors and collects input from Google Street View and video cameras inside the car.
- The AI simulates human perceptual and decision-making processes and controls actions in driver-control systems such as steering and brakes.
- The car's software consults Google Maps for advance notice of things like landmarks and traffic signs and lights.
- An override function is available to allow a human to take control of the vehicle.

The Pros and Cons of Autonomous Cars

The top benefit touted by autonomous vehicle proponents is safety. A U.S. Department of Transportation and NHTSA statistical projection of traffic fatalities for 2017 estimated that 37,150 people died in motor vehicle traffic crashes that year. The NHTSA estimates that 94% of serious crashes are due to human error or poor choices, such as drunk or distracted driving. Autonomous cars remove those risk factors from the equation -- though self-driving cars are still vulnerable to other factors, such as mechanical issues, that cause crashes.

If autonomous cars can significantly reduce the number of crashes, the economic benefits could be enormous. Injuries impact economic activity, including $57.6 billion in lost workplace productivity, and $594 billion due to loss of life and decreased quality of life due to injuries, according to the NHTSA.

In theory, if the roads were mostly occupied by autonomous cars, traffic would flow smoothly and there would be less traffic congestion. In cars that are fully automated, the occupants could do productive activities while commuting to work. People who aren't able to drive due to physical limitations could find new independence through autonomous vehicles and would have the opportunity to work in fields that require driving.

Autonomous trucks have been tested in the U.S. and Europe to allow drivers to use autopilot over long distances, freeing the driver to rest or complete tasks and improving driver safety and fuel efficiency. This initiative, called truck platooning, is powered by adaptive cruise control (ACC), collision avoidance systems and vehicle-to-vehicle communications for cooperative adaptive cruise control (CACC).

The downsides of autonomous vehicle technology could be that riding in a vehicle without a driver behind the steering wheel may be unnerving, at least at first. But as self-driving capabilities become commonplace, human drivers may become overly reliant on the autopilot technology and leave their safety in the hands of automation, even when they should act as backup drivers in case of software failures or mechanical issues.

In one example from March 2018, Tesla's Model X SUV was on autopilot when it crashed into a highway lane divider. The driver's hands were not on the wheel, despite visual warnings and an audible warning to put his hands back on the steering wheel, according to the company.

Autonomous Car Safety and Challenges

Autonomous cars must learn to identify countless objects in the vehicle's path, from branches and litter to animals and people. Other challenges on the road are tunnels that interfere with Global Positioning Systems (GPS), construction projects that cause lane changes, or complex decisions, like where to stop to allow emergency vehicles to pass.

The systems need to make instantaneous decisions on when to slow down, swerve or continue acceleration normally. This is a continuing challenge for developers and there are reports of self-driving cars hesitating and swerving unnecessarily when objects are detected in or near the roadways.

This problem was clear in a fatal accident in March 2018 which involved an autonomous car operated by Uber. The company reported that the vehicle's software identified a pedestrian but deemed it a false positive and failed to swerve to avoid hitting her. This crash caused Toyota to temporarily cease its testing of self-driving cars on public roads, but its testing will continue elsewhere. The Toyota Research Institute is constructing a test facility on a 60-acre site in Michigan to further develop automated vehicle technology.

With crashes also comes the question of liability, and lawmakers have yet to define who is liable when an autonomous car is involved in an accident. There are also serious concerns that the software used to operate autonomous vehicles can be hacked, and automotive companies are working to address cyber security risks.

Car makers are subject to Federal Motor Vehicle Safety Standards, and the NHTSA reports that more work must be done for vehicles to meet those standards.

The Road to Driverless Cars

The path toward self-driving cars began with incremental automation features for safety and convenience before the year 2000, with cruise control and antilock brakes. After the turn of the millennium, advanced safety features including electronic stability control, blind spot detection and collision and lane shift warnings became available in vehicles. Between 2010 and 2016, advanced driver assistance capabilities such as rearview video cameras, automatic emergency brakes and lane centering assistance emerged, according to the NHTSA.

Since 2016, automation has moved toward partial autonomy, with features that help drivers stay in their lane, along with adaptive cruise control technology, and the ability to self-park.

Fully automated vehicles are not publicly available yet and may not be for many years. In the U.S., the NHTSA provides federal guidance for introducing automated driving systems onto public roads and as autonomous car technologies advance, so will the department's guidance.

Self-driving cars are not yet legal on most roads. In June 2011, Nevada became the first jurisdiction in the world to allow driverless cars to be tested on public roadways; California, Florida, Ohio and Washington, D.C., have followed in the years since.

Autonomous Logistics

The modern logistics industry is becoming a hub for exciting robotic applications. It was slow to adopt automation at first, but the last few years have seen a rise in robotic solutions to logistics problems. According to Machine Handling and Logistics, "Some of the largest [logistics] operators in the US have plans to automate almost every physical move in their facilities within the next two to three years".

These applications range from (somewhat mundane) palletizing tasks all the way up to self-driving trucks and delivery drones.

Here are ten of the most interesting emerging applications in autonomous logistics today.

Warehousing Robots

One of the most mature robotic technologies in logistics is autonomous warehousing. Since Amazon acquired Kiva Systems in 2012, robotic warehouses have grown in popularity. They reduce the time that workers spend traveling around the warehouse by having robots bring shelves to the workers instead.

Although Amazon was one of the first big businesses on the scene, their robotic warehouses certainly aren't the only ones. In May 2018, logistics giant Rakuten Super Logistics partnered with InVia Robotics to automate their US warehouses, and more partnerships are surely on the way.

Autonomous Picking

Most warehousing robots these days work by bringing shelves to human workers, who then pick items by hand. However, more and more robots are able to pick items from static shelves without human assistance. This involves grasping diversely shaped items—a complex task. It would be great for logistics companies if all robots were capable of this, which is why Amazon hosts its annual Picking Challenge.

Robotic Packing

Packing is one of the most common logistics tasks—after all, everything shipped must be packed. Robots can be used for many of the several steps in packaging. At first, products are packed into individual packages. This is often done by the manufacturer. The next two steps are additional packing into larger boxes and crates; the latter step, called tertiary packaging, is often carried out by the logistics company.

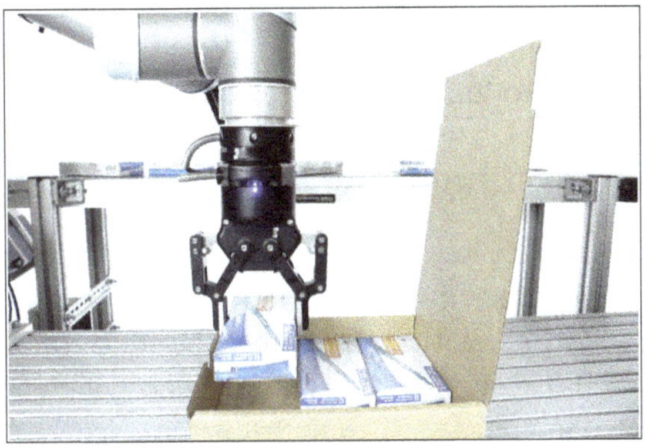

Packing is one of the most common tasks in logistics.

Robot Palletizing

One example of tertiary packaging is palletizing. It's well suited to robots because it generally involves repetitive actions and items that are too heavy for humans to lift.

There are even specialized palletizing robots: these are usually robotic arms that have five degrees of freedom (DoF) and a sixth DoF that can rotate more than 360 degrees, so that the robot can properly orient items onto pallets. There are also many programming solutions to simplify robot palletizing, such as the UR palletizing wizard.

Robotic Ports

One of the newest robotic-logistic applications is robotic ports. These aren't USB or Ethernet ports—they're where cargo is loaded onto boats.

Robotic ports were first heard about back in May when a Chinese port was made fully robotic. The port was designed by TuSimple, a self-driving-truck startup. It uses autonomous cranes and self-driving container vehicles to perform all the functions of a cargo port, but with no humans on site.

Self-driving Vehicles

With the rapid advances made in the past few years, autonomous trucks are the most obvious example of self-driving technology. Other applications include autonomous forklifts, delivery vehicles, and mobile parcel stations (like mailboxes on wheels).

As with the TuSimple port, China seems to be a driving force (if you'll excuse the pun) for autonomous logistics vehicles. "Dozens of autonomous truck startups are reported to have launched in China in the past two years".

Last Mile Drone Delivery

Last mile delivery is (as the name suggests) the last step in the logistics journey from manufacturer to end customer. Although drone delivery has received a lot of media attention, drones aren't the only robots being tested at this stage. Other recent developments include autonomous ground, super-size, and four-legged delivery robots.

Container Loading and Unloading

Container loading is like palletizing, but instead of piling products onto a pallet, it involves pulling packages out of a shipping container. As a result, the handling requirements are slightly different, since the robot must be able to hold products from the side. One example of such a system is the IRIS project, which seeks to use vision and self-learning to improve packing.

Sorting

Sorting is essential at many stages of logistics. But it has traditionally been challenging for robots, which require advanced sensing to differentiate between various products. However, the necessary capabilities are easier to acquire than ever before thanks to vision sensors like the Robotiq Wrist Camera.

Collaborative Robot Packing

As you know if you've been following the Robotiq blog, cobots are on the rise in many industries, and logistics is no different. Collaborative packing lets humans and cobots work together: the robot can take over the more mundane, repetitive actions, while humans handle the more delicate tasks.

Autonomous Spaceport Drone Ship

SpaceX recently managed to land its Falcon 9 reusable rocket on its autonomous drone ship out at sea after four unsuccessful attempts, with plans for at least two more sea landings later this year. It takes a special kind of ship to catch a rocket falling from the sky, and here's how they do it.

Elon Musk just had the best week of business in his life after finally successfully landing the SpaceX Falcon 9 rocket engine on a tiny speck of a refitted cargo barge floating in the ocean, in addition to the 325,000 reservations for the Tesla Model 3 electric sedan.

But the riskiest venture of the week was the final moment the Falcon 9 rocket engine returned to Earth after launching an addition to the International Space Station into orbit and, for the first time, managed to land upright on SpaceX's autonomous landing ship, the wonderfully-named "Of Course I Still Love You".

It turns out that the landing platforms that SpaceX has developed are just as complicated as literal rocket science.

There have been three separate landing ships developed by SpaceX. The concept of an "Autonomous Spaceport Drone Ship" harks back to Elon Musk's plans for "creating a paradigm shift in the traditional approach for reusing rocket hardware" in 2009, but the contract for a refitted barge was officially announced in 2014.

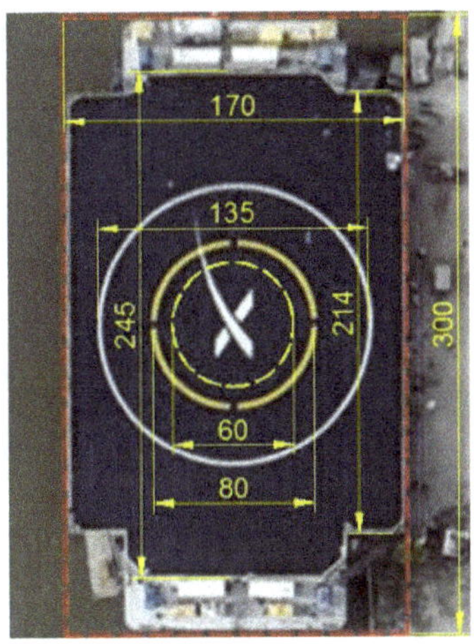

Photo originally tweeted (and since deleted) by Elon Musk of the landing platform.

The large, flat surface of the ocean barge and the ability to transport heavy payloads are necessary for the high-velocity landings of the Faclon 9 rocket. SpaceX refits the barge platforms with an extended platform measuring 170 feet by 300 feet to facilitate the 18 foot span of the landing legs and the rocket's support supplies. The first ship featured extended wings and blast walls, but the current models no longer have these features.

According to NASA, these Marmac autonomous drone ships are outfitted with four diesel-powered azimuth thruster engines which can pivot horizontally, removing the need for a ship rudder and offering better maneuverability with a complete 360 degree range of motion for the ship controls. It was originally reported that these thrusters help hold the drone ship position within three meters of accuracy. The engines are supplied by marine equipment manufacturer Thrustmaster.

The materials on the surface of the ship aren't entirely known, but is likely a thick sheet of steel which prominently features the SpaceX logo.

The drone ships are capable of maneuvering autonomously, using GPS information for precision positioning, but they can also be remotely controlled by an accompanying support ship with a crew of technicians standing by.

A wide array of sensors are used in tandem with the GPS information to manage the "attitude and placement" of the drone ship and communicate with the incoming rocket and the onlooking teams to control and coordinate the complicated precise landings.

Following each landing attempt the crews of the support ship board the drone ship. If the landing is successful, the standing rocket is welded to the deck of the ship and reinforced until it arrives back in port. If the landing was a failure, whatever is salvageable is secured and returned to port.

The Atlantic drone ship returns to the Port of Jacksonville or Port Canaveral in Florida, and the Pacific drone ship docks in the Port of Los Angeles in California.

The first "Just Read The Instructions," which was retired after six months of service. Photo: SpaceX.

The "Of Course I Still Love You," which received the first successful Falcon 9 landing April 8th, 2016. Photo: SpaceX .

The first implemented use of SpaceX's Autonomous Spaceport Drone Ship was in the Atlantic Ocean following a cargo launch to the International Space Station for NASA on January 10th, 2015, which resulted in a failed landing attempt.

Elon Musk claims the liklihood of a successful landing was 50 percent at the most, but recently stated that the odds are closer to 30 percent.

The first drone ship was based on a Marmac series 300 ocean barge, and its name was revealed to be "Just Read The Instructions" after its first test landing.

The first drone ship "Just Read The Instructions" was replaced for Atlantic Ocean duties after six months of operations and two failed test landings at sea.

It's replacement platform was leased by SpaceX and developed on a much newer Marmac barge, model 304, and as I mentioned earlier, named "Of Course I Still Love You."

Both ship titles are creatively named after autonomously-controlled spaceships from the novel The Player of Games.

A third floating landing platform based on the leased Marmac barge model number 303 was finished in the spring of 2015 and transited the Panama Canal for Pacific Ocean rocket recovery duties. It was later named “Just Read The Instructions” like the first, now retired floating landing platform.

SpaceX has attempted to land the Falcon 9 rocket engine on an autonomous drone ship five times. Two landing attempts in January and April of 2015 resulted in failures after contact with the drone ship, but many stages of the attempts were successful.

The third intended landing attempt was thwarted in June 2015 after the rocket disintegrated before entering the first stage of the landing process.

The first Pacific landing attempting, and fourth total, was conducted in January with a successful “soft landing” of the rocket on the drone ship. However, a catch pin in one of the rocket’s landing legs failed and the rocket toppled over.

The fifth landing attempt on Friday was the first full success in SpaceX’s attempts to land its Falcon 9 rocket engine on a drone ship at sea. This success, and the hope for many more like it in the future, mark the underlying purpose behind SpaceX’s reusable rocket program.

The ability to completely recover an entire rocket body and return it to port to prep for reuse will drastically cut the cost of spaceflight and increase the pace at which payloads and, eventually, humans will be able to access space.

According to Elon Musk at SpaceX’s press conference following the successful landing, eventual use of the drone ships should drop to a quarter of SpaceX return landings with a shift towards land locations for returning rockets. Elon Musk also once tweeted an eventual possibility where the rocket engines land on the drone ship, are prepped and refueled in a rapid turnaround time, and then fly themselves “home”.

Musk also brought up plans for the first manned SpaceX flight by the end of next year with the second generation of the Dragon capsule, with an unmanned test planned first.

“We’ll be successful when it’s boring,” Musk quipped at a question asking him about the next goal following this major success. But he has a point—the realization that this sort of marveling engineering achievement brings us one step closer to casual, inexpensive spaceflight is truly sobering.

The ability for a rocket delivering payloads beyond our atmosphere to communicate and successfully connect with an autonomous floating platform in rocky seas not only breaks the traditional mold of spaceflight recovery, but almost literally jumps from the pages of science fiction.

Cyborg

A cybernetic organism or "cyborg" in IT is defined as an organism with both biological and technological components. In some definitions, a cyborg is described as a hypothetical or fictional creation. However, in a technical sense, humans can be seen as cyborgs in various types of situations, including the use of artificial implants.

Part of the diverse use of the word "cyborg" revolves around how humans see their interactions with technology. A person could be considered a cyborg when they are outfitted with implants such as artificial heart valves, cochlear implants or insulin pumps. A person could even be called a cyborg when they are using specific wearable technologies like Google Glass, or even using laptops or mobile devices to do work.

However, a different definition of a cyborg involves fictional pictures of human individuals with enhanced virtual-reality vision, robotic implants on limbs and torso, and other more significant body IT components. The popular definition of cyborg changes as a range of science-fiction-type ideas become realities.

Laboratory Automation

Laboratory testing has grown from a manual, "hands-on" process providing a simple test menu to an instrument-centric, high-volume clinical engine inside the modern healthcare enterprise. While the instruments have grown both in size and scale, the proverbial "glue" that holds the laboratory together is the lines of automation. These "conveyor belts" for laboratory samples bring fast and accurate routing of specimens to specific points in the laboratory work flow. Because of this growth, automation has moved past the "nice to have" to the "must have" in the modern laboratory, and automation solutions are beginning to extend their footprints into other areas of the laboratory such as microbiology and molecular diagnostics.

With the three components of automation growing within the clinical laboratory in the late 1980s (instrument-level automation, the LIS/LIMS, and early pre- and post-analytic automation), the strategies for laboratory automation have taken a variety of routes. First, the degree of automation at the level of the instrument has grown to handle higher throughputs as the demands of the laboratory have grown. This growth is easily seen in the orders of magnitude increases in instrument throughput over the past few decades. The LIS/LIMS systems have grown from simple appliances that handle the mechanics of laboratory billing to complete, end-to-end platforms that govern the entire business process and work flow of the laboratory. These IT systems are now mission-critical to the operation of a laboratory and must work in concert with instruments and automation for a laboratory to be successful.

The pre- and post-analytic automation solutions are present in one of two varieties: open and closed. Of the two, closed automation is the more common. Closed automation solutions are those that are provided by instrument manufacturers and typically connect only to instruments from that vendor. And unless a single vendor has instruments in multiple parts of the lab, closed automation solutions exist as islands of automation around their specific parts of the lab. Open automation solutions exist independently from the instruments in the laboratory. These solutions are built by independent companies and interface to the instruments and the LIS/LIMS to automate the pre- and post-analytic work flows.

The main difference between the open and closed approaches is the types of instruments connected to the automation line. Unlike the closed lines that are provided with instruments, the open automation solutions are designed and acquired by labs independently of the instruments in the laboratory. While the open automation lines connect to instruments in a similar fashion (direct track sampling or robotic arm), open automation lines can cross service areas of the lab because of the ability to interface to any instrument, regardless of the vendor. This independence from a specific vendor allows open automation systems to transport specimens across various parts of the lab and provide a single automation solution for the entire laboratory.

The current "whys" for automation in the laboratory are error reduction and staff augmentation. Both are driven by increased testing demands and laboratory staffing needs. Automation has indeed moved from "nice to have" for large reference laboratories to "must have" for any clinical laboratory. The current climate for reimbursements for laboratory testing has only increased the importance of automation, with further reductions that push laboratories to optimize laboratory throughput and staffing to keep up with the increasing demands.

Benefits of Lab Automation

Provides Valuable Walk-away Time

There are many important tasks to get done in the lab, so why not eliminate the tedious and time consuming task of hand labeling tubes? Imagine how much time you will save by giving this job to a fully automated labeling machine that can fill, read 2D barcode for verification, and place your tubes back into the rack. Labeling tubes is a rather simple task, so is it really necessary to have a scientist perform such tedious work when he or she could be using their valuable skills for a more specialized task.

Reduces Repetitive Injuries

Manual labeling of tubes is not only tedious and time-consuming but it can also lead to physical strain of the fingers and hands. People who have been hand labeling tubes for years might start to experience numbness and tingling in the hand, which ultimately decreases productivity in the lab. If your daily routine consists of labeling, filling, uncapping and capping of tubes, this can cause stress injuries in the fingers and in many cases lead to carpal tunnel syndrome. Many fully-automated labeling machines can complete all of these tasks for you and decrease the risk of repetitive injuries from manual labeling.

Reduces Costs in the Lab

Using laboratory automation in your lab will decrease the number of lab technicians needed to label tubes, therefore resulting in cost savings. Not only will you reduce costs in the lab, but you also might find an increase in revenue due to lab automation.

Eliminates Human Error

There are many steps involved in preparing and keeping track of samples in the lab. Lab automation not only decreases human error associated with sample preparation but also it increases efficiency in the lab by allowing the user to track samples. If you are working with a LIMS system or excel work list, these can easily be imported into the software to ensure tracking of samples, therefore reducing the amount of lost or misplaced samples.

Flexible Machinery Saves Valuable Space

Saving space in the lab is a legitimate concern. Although some lab automation systems have a large footprint and take up lots of space, there are also many benchtop systems that are designed to fit easily in a lab work space.

Laboratory Robotics

Laboratory robotics is the practice of using robots to perform or assist in laboratory tasks. While laboratory robots have found applications in diverse industries and sciences, pharmaceutical companies have used them more than any other group.

Advantages of laboratory robotics include high speed, high efficiency, minimal wastage, task reproducibility, task endurance, precision, high productivity, enhanced safety for laboratory personnel, the ability to withstand adverse environmental conditions, and reduced tedium and boredom among laboratory workers who would otherwise perform the tasks.

Laboratory robotics has some shortcomings and limitations such as high initial cost, limited flexibility, and lack of intuition for solving difficult or esoteric problems. In addition, layoffs of human workers (as robots replace them) can damage employee morale.

References

- Spacexs-landing-drone-ship-is-just-as-complicated-as-th-1769987148: foxtrotalpha.jalopnik.com, Retrieved 10 April 2020
- Advantages-and-disadvantages-of-automation, technology: britannica.com, Retrieved 16 June 2020
- 5-benefits-lab-automation: scinomix.com, Retrieved 09 July 2020
- Automation, technology: britannica.com, Retrieved 17 April 2020
- Laboratory-robotics: searchenterpriseai.techtarget.com, Retrieved 27 May 2020
- Driverless-car: searchenterpriseai.techtarget.com, Retrieved 28 May 2020
- Cyborg-15651: techopedia.com, Retrieved 30 June 2020

6

Applications of Robotics

Bionics is the branch of science which applies the biological methods and systems found in nature to the study and design of engineering systems and modern technology. One such application of robotics is the shadow hand which is a humanoid robot hand system and is comparable to a human hand in size and shape. The diverse applications of robotics in the current scenario have been thoroughly discussed in this chapter.

Bionics

Bionics is the application of biological methods and systems found in nature to the study and design of engineering systems and modern technology.

The word *bionic* was coined by Jack E. Steele in 1958, possibly originating from the technical term *bion* (pronounced *BEE-on*; from Ancient Greek: βίος), meaning 'unit of life' and the suffix *-ic*, meaning 'like' or 'in the manner of', hence 'like life'. Some dictionaries, however, explain the word as being formed as a portmanteau from *biology* and *electronics*. It was popularized by the 1970s U.S. television series *The Six Million Dollar Man* and *The Bionic Woman*, both based upon the novel *Cyborg* by Martin Caidin, which was itself influenced by Steele's work. All feature humans given superhuman powers by electromechanical implants.

The transfer of technology between lifeforms and manufactures is, according to proponents of bionic technology, desirable because evolutionary pressure typically forces living organisms, including fauna and flora, to become highly optimized and efficient. A classical example is the development of dirt- and water-repellent paint (coating) from the observation that the surface of the lotus flower plant is practically unsticky for anything (the lotus effect).

The term "biomimetic" is preferred when reference is made to chemical reactions. In that domain, biomimetic chemistry refers to reactions that, in nature, involve biological macromolecules (e.g. enzymes or nucleic acids) whose chemistry can be replicated *in vitro* using much smaller molecules.

Examples of bionics in engineering include the hulls of boats imitating the thick skin of dolphins; sonar, radar, and medical ultrasound imaging imitating animal echolocation.

In the field of computer science, the study of bionics has produced artificial neurons, artificial neural networks, and swarm intelligence. Evolutionary computation was also motivated by bionics ideas but it took the idea further by simulating evolution in silico and producing well-optimized solutions that had never appeared in nature.

It is estimated by Julian Vincent, professor of biomimetics at the University of Bath's Department of Mechanical Engineering, that "at present there is only a 12% overlap between biology and technology in terms of the mechanisms used".

History

The name biomimetics was coined by Otto Schmitt in the 1950s. The term bionics was coined by Jack E. Steele in 1958 while working at the *Aeronautics Division House* at Wright-Patterson Air Force Base in Dayton, Ohio. However, terms like biomimicry or biomimetics are more preferred in the technology world in efforts to avoid confusion between the medical term bionics. Coincidentally, Martin Caidin used the word for his 1972 novel *Cyborg*, which inspired the series *The Six Million Dollar Man*. Caidin was a long-time aviation industry writer before turning to fiction full-time.

Methods

Velcro was inspired by the tiny hooks found on the surface of burs.

Often, the study of bionics emphasizes implementing a function found in nature rather than just imitating biological structures. For example, in computer science, cybernetics tries to model the feedback and control mechanisms that are inherent in intelligent behavior, while artificial intelligence tries to model the intelligent function regardless of the particular way it can be achieved.

The conscious copying of examples and mechanisms from natural organisms and ecologies is a form of applied case-based reasoning, treating nature itself as a database of solutions that already work. Proponents argue that the selective pressure placed on all natural life forms minimizes and removes failures.

Although almost all engineering could be said to be a form of biomimicry, the modern origins of this field are usually attributed to Buckminster Fuller and its later codification as a house or field of study to Janine Benyus.

Roughly, we can distinguish three biological levels in the fauna or flora, after which technology can be modeled:

- Mimicking natural methods of manufacture.
- Imitating mechanisms found in nature (velcro).
- Studying organizational principles from the social behaviour of organisms, such as the flocking behaviour of birds, optimization of ant foraging and bee foraging, and the swarm intelligence (SI)-based behaviour of a school of fish.

Examples

- In robotics, bionics and biomimetics are used to apply the way animals move to the design of robots. BionicKangaroo was based on the movements and physiology of kangaroos.
- Velcro is the most famous example of biomimetics. In 1948, the Swiss engineer George de Mestral was cleaning his dog of burrs picked up on a walk when he realized how the hooks of the burrs clung to the fur.
- The horn-shaped, saw-tooth design for lumberjack blades used at the turn of the 19th century to cut down trees when it was still done by hand was modeled after observations of a wood-burrowing beetle. It revolutionized the industry because the blades worked so much faster at felling trees.
- Cat's eye reflectors were invented by Percy Shaw in 1935 after studying the mechanism of cat eyes. He had found that cats had a system of reflecting cells, known as tapetum lucidum, which was capable of reflecting the tiniest bit of light.
- Leonardo da Vinci's flying machines and ships are early examples of drawing from nature in engineering.
- Resilin is a replacement for rubber that has been created by studying the material also found in arthropods.
- Julian Vincent drew from the study of pinecones when he developed in 2004 "smart" clothing that adapts to changing temperatures. "I wanted a nonliving system which would respond to changes in moisture by changing shape", he said. "There are several such systems in plants, but most are very small – the pinecone is the largest and therefore the easiest to work on". Pinecones respond to higher humidity by opening their scales (to disperse their seeds). The "smart" fabric does the same thing, opening up when the wearer is warm and sweating, and shutting tight when cold.

- "Morphing aircraft wings" that change shape according to the speed and duration of flight were designed in 2004 by biomimetic scientists from Penn State University. The morphing wings were inspired by different bird species that have differently shaped wings according to the speed at which they fly. In order to change the shape and underlying structure of the aircraft wings, the researchers needed to make the overlying skin also be able to change, which their design does by covering the wings with fish-inspired scales that could slide over each other. In some respects this is a refinement of the swing-wing design.

Lotus leaf surface, rendered: microscopic view.

- Some paints and roof tiles have been engineered to be self-cleaning by copying the mechanism from the Nelumbo lotus.
- Cholesteric liquid crystals (CLCs) are the thin-film material often used to fabricate fish tank thermometers or mood rings, that change color with temperature changes. They change color because their molecules are arranged in a helical or chiral arrangement and with temperature the pitch of that helical structure changes, reflecting different wavelengths of light. Chiral Photonics, Inc. has abstracted the self-assembled structure of the organic CLCs to produce analogous optical devices using tiny lengths of inorganic, twisted glass fiber.
- Nanostructures and physical mechanisms that produce the shining color of butterfly wings were reproduced in silico by Greg Parker, professor of Electronics and Computer Science at the University of Southampton and research student Luca Plattner in the field of photonics, which is electronics using photons as the information carrier instead of electrons.
- The wing structure of the blue morpho butterfly was studied and the way it reflects light was mimicked to create an RFID tag that can be read through water and on metal.
- The wing structure of butterflies has also inspired the creation of new nanosensors to detect explosives.

- Neuromorphic chips, silicon retinae or cochleae, has wiring that is modelled after real neural networks. *S.a.:* connectivity.
- Technoecosystems or 'EcoCyborg' systems involve the coupling of natural ecological processes to technological ones which mimic ecological functions. This results in the creation of a self-regulating hybrid system. Research into this field was initiated by Howard T. Odum, who perceived the structure and emergy dynamics of ecosystems as being analogous to energy flow between components of an electrical circuit.
- Medical adhesives involving glue and tiny nano-hairs are being developed based on the physical structures found in the feet of geckos.
- Computer viruses also show similarities with biological viruses in their way to curb program-oriented information towards self-reproduction and dissemination.
- The cooling system of the Eastgate Centre building, in Harare was modeled after a termite mound to achieve very efficient passive cooling.
- Adhesive which allows mussels to stick to rocks, piers and boat hulls inspired bioadhesive gel for blood vessels.
- Through the field of bionics, new aircraft designs with far greater agility and other advantages may be created. This has been described by Geoff Spedding and Anders Hedenström in an article in Journal of Experimental Biology. Similar statements were also made by John Videler and Eize Stamhuis in their book Avian Flight and in the article they present in Science about LEVs. John Videler and Eize Stamhuis have since worked out real-life improvements to airplane wings, using bionics research. This research in bionics may also be used to create more efficient helicopters or miniature UAVs. This latter was stated by Bret Tobalske in an article in Science about Hummingbirds. Bret Tobalske has thus now started work on creating these miniature UAVs which may be used for espionage. UC Berkeley as well as ESA have finally also been working in a similar direction and created the Robofly (a miniature UAV) and the Entomopter (a UAV which can walk, crawl and fly).

Specific Uses of the Term

In Medicine

Bionics is a term which refers to the flow of concepts from biology to engineering and vice versa. Hence, there are two slightly different points of view regarding the meaning of the word.

In medicine, bionics means the replacement or enhancement of organs or other body parts by mechanical versions. Bionic implants differ from mere prostheses by mimicking the original function very closely, or even surpassing it.

Bionics' German equivalent, *Bionik*, always adheres to the broader meaning, in that it tries to develop engineering solutions from biological models. This approach is motivated by the fact that biological solutions will usually be optimized by evolutionary forces.

While the technologies that make bionic implants possible are still in infancy, a few bionic items already exist, the best known being the cochlear implant, a device for deaf people. By 2004 fully functional artificial hearts were developed. Significant progress is expected with the advent of nanotechnology. A well-known example of a proposed nanodevice is a respirocyte, an artificial red cell, designed (though not built yet) by Robert Freitas.

Kwabena Boahen from Ghana was a professor in the Department of Bioengineering at the University of Pennsylvania. During his eight years at Penn, he developed a silicon retina that was able to process images in the same manner as a living retina. He confirmed the results by comparing the electrical signals from his silicon retina to the electrical signals produced by a salamander eye while the two retinas were looking at the same image.

In 2007 the Scottish company Touch Bionics launched the first commercially available bionic hand, named "i-Limb Hand". According to the firm, by May 2010 it has been fitted to more than 1,200 patients worldwide.

The Nichi-In group is working on biomimicking scaffolds in tissue engineering, stem cells and regenerative medicine have given a detailed classification on biomimetics in medicine.

On 21 July 2015, the BBC's medical correspondent Fergus Walsh reported, "Surgeons in Manchester have performed the first bionic eye implant in a patient with the most common cause of sight loss in the developed world. Ray Flynn, 80, has dry age-related macular degeneration which has led to the total loss of his central vision. He is using a retinal implant which converts video images from a miniature video camera worn on his glasses. He can now make out the direction of white lines on a computer screen using the retinal implant." The implant, known as the Argus II and manufactured in the US by the company Second Sight Medical Products, had been used previously in patients who were blind as the result of the rare inherited degenerative eye disease retinitis pigmentosa.

Politics

A political form of biomimicry is bioregional democracy, wherein political borders conform to natural ecoregions rather than human cultures or the outcomes of prior conflicts.

Critics of these approaches often argue that ecological selection itself is a poor model of minimizing manufacturing complexity or conflict, and that the free market relies on conscious cooperation, agreement, and standards as much as on efficiency – more analogous to sexual selection. Charles Darwin himself contended that both were balanced in natural selection – although his contemporaries often avoided frank talk about sex, or any suggestion that free market success was based on persuasion, not value.

Advocates, especially in the anti-globalization movement, argue that the mating-like processes of standardization, financing and marketing, are already examples of runaway evolution – rendering a system that appeals to the consumer but which is inefficient at use of energy and raw materials. Biomimicry, they argue, is an effective strategy to restore basic efficiency.

Biomimicry is also the second principle of Natural Capitalism.

Other Uses

Business biomimetics is the latest development in the application of biomimetics. Specifically it applies principles and practice from biological systems to business strategy, process, organisation design and strategic thinking. It has been successfully used by a range of industries in FMCG, defence, central government, packaging and business services. Based on the work by Phil Richardson at the University of Bath the approach was launched at the House of Lords in May 2009.

In a more specific meaning, it is a creativity technique that tries to use biological prototypes to get ideas for engineering solutions. This approach is motivated by the fact that biological organisms and their organs have been well optimized by evolution. In chemistry, a biomimetic synthesis is a chemical synthesis inspired by biochemical processes.

Another, more recent meaning of the term bionics refers to merging organism and machine. This approach results in a hybrid system combining biological and engineering parts, which can also be referred as a cybernetic organism (cyborg). Practical realization of this was demonstrated in Kevin Warwick's implant experiments bringing about ultrasound input via his own nervous system.

Shadow Hand

The Shadow Dexterous Hand is a humaniform (humanoid) robot hand system developed by The Shadow Robot Company in London. The hand is comparable to a human hand in size and shape, and reproduces all of its degrees of freedom. The Hand is commercially available in pneumatic- and electric-actuated models and currently used in a wide range of institutions including NASA, Bielefeld University and Carnegie Mellon University, and EU research projects such as HANDLE.

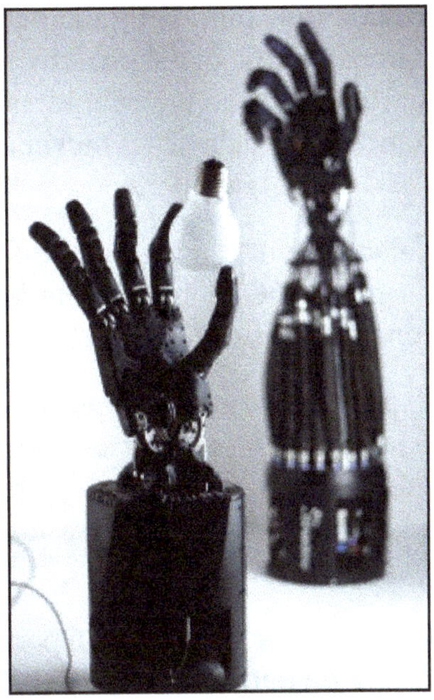

The Shadow C6M Smart Motor Hand in front of the Shadow C3 Dexterous Air Muscle Hand.

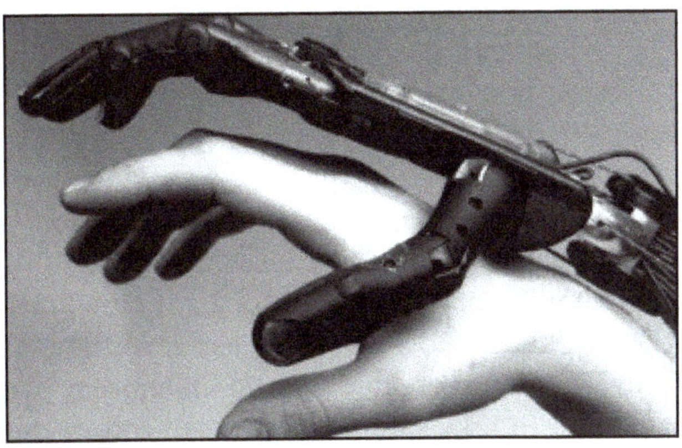

Comparison of The Shadow Dexterous Hand with the human hand.

The Shadow Dexterous Robot Hand is the first commercially available robot hand from the company, and follows a series of prototype humanoid hand and arm systems.

Design

The Shadow Dexterous Hand has been designed to be similar to the average hand of a human male. The forearm structure is slightly wider than a human forearm.

The Shadow Dexterous Hand has 24 joints. It has 20 degrees of freedom, greater than that of a human hand. It has been designed to have a range of movement equivalent to that of a typical human being. The four fingers of the hand contain two one-axis joints connecting the distal phalanx, middle phalanx and proximal phalanx and one universal joint connecting the finger to the metacarpal. The little finger has an extra one-axis joint on the metacarpal to provide the Hand with a palm curl movement. The thumb contains one one-axis joint connecting the distal phalanx to the proximal phalanx, one universal joint connecting the thumb to the metacarpal and one one-axis joint on the bottom of the metacarpal to provide a palm curl movement. The wrist contains two joints, providing flex/extend and adduct/abduct.

The hand is available in both electric motor driven and pneumatic muscle driven models. The motor hand is driven by 20 DC motors in the forearm, whereas the muscle hand is powered by 20 antagonistic pairs of Air Muscles in the forearm.

All hands have Hall effect sensors integrated into every joint to provide precise positional feedback. The motor hand includes force sensors for each degree of freedom and the muscle hand includes pressure sensors for each muscle. There are also several options for tactile sensing on the hand from basic pressure sensors to the BioTac multimodal tactile sensor from Syntouch LLC.

The Shadow Hand software system is based on Robot Operating System, through which configuration, calibration, simulation and control of the hand is implemented. A simulation of the Shadow hand can be downloaded and installed in ROS.

Rover

A rover (or sometimes planetary rover) is a space exploration vehicle designed to move across the surface of a planet or other celestial body. Some rovers have been designed to transport members of a human spaceflight crew; others have been partially or fully autonomous robots. Rovers usually arrive at the planetary surface on a lander-style spacecraft. Rovers are created to land on another planet, besides Earth, to find out information and to take samples. They can collect dust, rocks, and even take pictures.

Three different Mars rover designs; Sojourner, MER and Curiosity.

Comparison with Space Probes of Other Types

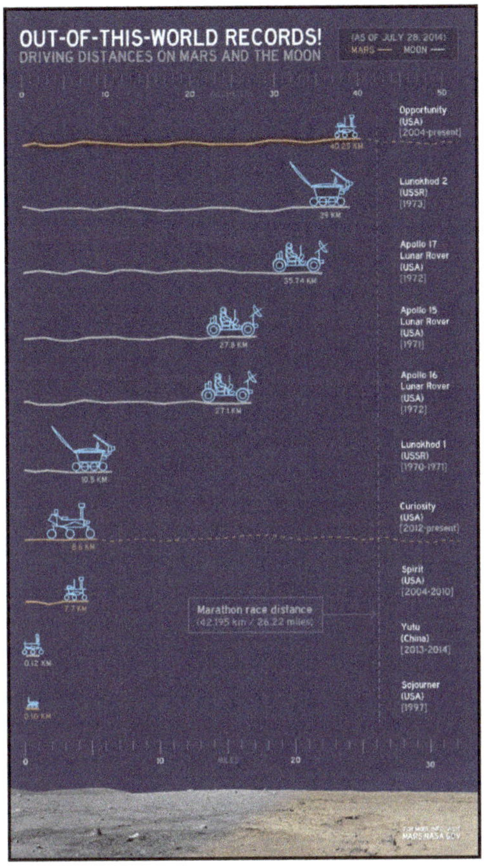

Comparison of distances driven by various wheeled vehicles on the surface of Earth's moon a nd Mars (NASA, 15 May 2013; updated version of 28 Jul 2014).

Rovers have several advantages over stationary landers: they examine more territory and they can be directed to interesting features. If they are solar powered, they can place themselves in sunny positions to weather winter months. They can also advance the knowledge of how to perform very remote robotic vehicle control which is necessarily semi-autonomous due to the finite speed of light.

Their advantages over orbiting spacecraft are that they can make observations to a microscopic level and can conduct physical experimentation. Disadvantages of rovers compared to orbiters are the higher chance of failure, due to landing and other risks, and that they are limited to a small area around a landing site which itself is only approximately anticipated.

Features

Rovers arrive on spacecraft and are used in conditions very distinct from those on the Earth, which makes some demands on their design.

Reliability

Rovers have to withstand high levels of acceleration, high and low temperatures, pressure, dust, corrosion, cosmic rays, remaining functional without repair for a needed period of time.

Mars rover Sojourner in cruise configuration.

Compactness

Rovers are usually packed for placing in a spacecraft, because it has limited capacity, and has to be deployed. They are also attached to a spacecraft, so devices for removing these connections are installed.

Autonomy

Rovers which land on celestial bodies far from the Earth, such as the Mars Exploration Rovers, cannot be remotely controlled in real-time since the speed at which radio signals travel is far too slow for *real time* or *near-real time* communication. For example, sending a signal from Mars to Earth takes between 3 and 21 minutes. These rovers are thus capable of operating autonomously with little assistance from ground control as far as navigation and data acquisition are concerned, although they still require human input for identifying promising targets in the distance to which to drive, and determining how to position itself to maximize solar energy. Giving a rover some rudimentary visual identification capabilities to make simple distinctions can allow engineers to speed up the reconnaissance.

Non-wheeled Approaches

Other rover designs that do not use wheeled approaches are possible. Mechanisms that utilize "walking" on robotic legs, hopping, rolling, etc. are possible. For example, Stanford University researchers have proposed "*Hedgehog*", a small cube-shaped rover that can controllably hop—or even spin out of a sandy sinkhole by corkscrewing upward to escape—for surface exploration of low gravity celestial bodies.

History

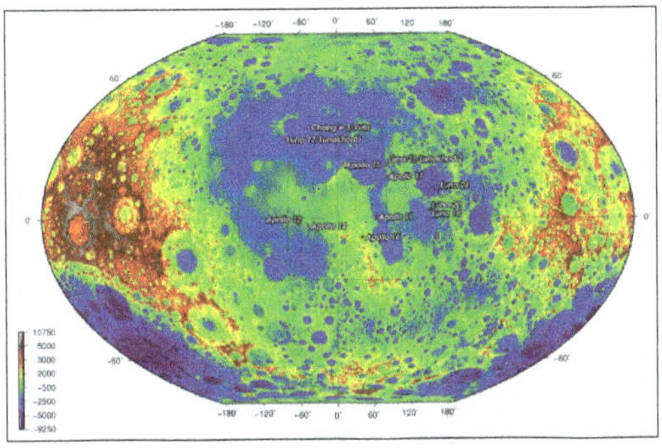

Landing sites of sample return and rover missions.

Lunokhod 0 (No.201)

The Soviet rover was intended to be the first roving remote-controlled robot on the Moon, but crashed during a failed start of the launcher 19 February 1969.

Lunokhod 1

The Lunokhod 1 rover landed on the Moon in November 1970. It was the first roving remote-controlled robot to land on any celestial body. The Soviet Union launched Lunokhod 1 aboard the Luna 17 spacecraft on November 10, 1970, and it entered lunar orbit on November 15. The spacecraft soft-landed in the Sea of Rains region on November 17. The lander had dual ramps from which Lunokhod 1 could descend to the lunar surface, which it did at 06:28 UT. From November 17, 1970 to November 22, 1970 the rover drove 197 m, and during 10 communication sessions returned 14 close up pictures of the Moon and 12 panoramic views. It also analyzed the lunar soil. The last successful communications session with Lunokhod 1 was on September 14, 1971. Having worked for 11 months, Lunokhod 1 held the durability record for space rovers for more than 30 years, until a new record was set by the Mars Exploration Rovers.

Apollo 15 Lunar Rover.

Apollo Lunar Roving Vehicle

NASA included Lunar Roving Vehicles in three Apollo missions: Apollo 15 (which landed on the Moon July 30, 1971), Apollo 16 (which landed April 21, 1972) and Apollo 17 (which landed December 11, 1972).

Lunokhod 2

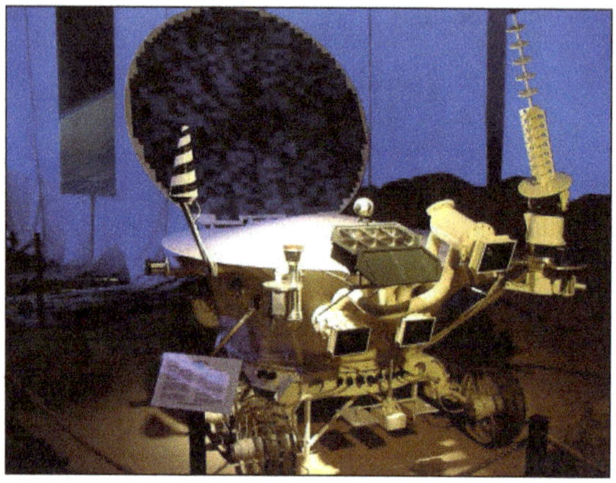

The Lunokhod 2 Lunar Rover.

The Lunokhod 2 was the second of two unmanned lunar rovers landed on the Moon by the Soviet Union as part of the Lunokhod program. The rover became operational on the Moon on January 16, 1973. It was the second roving remote-controlled robot to land on any celestial body. The Soviet Union launched Lunokhod 2 aboard the Luna 21 spacecraft on January 8, 1973, and it entered lunar orbit on January 12, 1973. The spacecraft soft-landed in the eastern edge of the Mare Serenitatis region on January 15, 1973. Lunokhod 2 descended from the lander's dual ramps to the lunar surface at 01:14 UT on January 16, 1973. Lunokhod 2 operated for about 4 months, covered 39 km (24 mi) of terrain, including hilly upland areas and rilles, and sent back 86 panoramic images and over 80,000 TV pictures. Based on wheel rotations Lunokhod 2 was thought to have covered 37 km (23 mi) but Russian scientists at the Moscow State University of Geodesy and Cartography (MIIGAiK) have revised that to an estimated distance of about 42.1–42.2 km (26.2–26.2 mi) based on Lunar Reconnaissance Orbiter (LRO) images of the lunar surface. Subsequent discussions with their American counterparts ended with an agreed-upon final distance of 39 km (24 mi), which has stuck since.

Prop-M Rover

The Soviet Mars 2 and Mars 3 landers had a small 4.5 kg Mars rover on board, which would have moved across the surface on skis while connected to the lander with a 15-meter umbilical. Two small metal rods were used for autonomous obstacle avoidance, as radio signals from Earth would have taken too long to drive the rovers using remote control. The rover was planned to be placed on the surface after landing by a manipulator arm and to move in the field of view of the television cameras and stop to make measurements every 1.5 meters. The rover tracks in the Martian soil would also have been recorded to determine material properties. Because of the crash landing of Mars 2 and the communication failure (15 seconds post landing) of Mars 3, neither rover was deployed.

Prop-M Rover.

Lunokhod 3

The Soviet rover was intended to be the third roving remote-controlled robot on the Moon in 1977. The mission was canceled due to lack of launcher availability and funding, although the rover was built.

Marsokhod

The Marsokhod was a *heavy* Soviet rover (hybrid, with both controls telecommand and automatic) aimed at Mars, part of the Mars 4NM and scheduled to be released (after 1973 according to the plans of 1970) launched by a N1 rocket that never arrived to fly successfully.

Sojourner on Mars.

Sojourner

The Mars Pathfinder mission included *Sojourner*, the first rover to successfully deploy on another planet. NASA, the space agency of the United States, launched Mars Pathfinder on 4th December 1996; it landed on Mars in a region called Chryse Planitia on 4th July 1997. From its landing until the final data transmission on 27th September 1997, Mars Pathfinder returned 16,500 images from the lander and 550 images from *Sojourner*, as well as data from more than 15 chemical analyses of rocks and soil and extensive data on winds and other weather factors.

Beagle 2 Planetary Undersurface Tool

Beagle 2 was designed to explore Mars with a small "mole" (Planetary Undersurface Tool, or PLUTO), to be deployed by the arm. PLUTO had a compressed spring mechanism designed to enable it to move across the surface at a rate of 20 mm per second and to burrow into the ground and collect a subsurface sample in a cavity in its tip. Beagle 2 failed while attempting to land on Mars in 2003.

Mars Exploration Rover A *Spirit*

Spirit is a robotic rover on Mars, active from 2004 to 2010. It was one of two rovers of NASA's ongoing Mars Exploration Rover Mission. It landed successfully on Mars at 04:35 Ground UTC on January 4, 2004, three weeks before its twin, *Opportunity* (MER-B), landed on the other side of the planet. Its name was chosen through a NASA-sponsored student essay competition. The rover became stuck in late 2009, and its last communication with Earth was sent on March 22, 2010.

Active Rover Missions

Active Rover Locations in Context

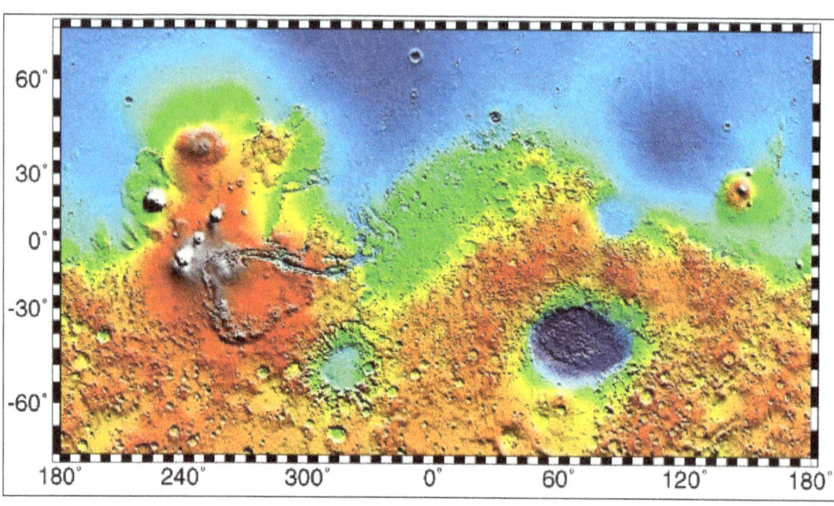

Interactive imagemap of the global topography of Mars, overlain with locations of Mars landers and rovers. Hover your mouse to see the names of prominent geographic features, and click to link to them. Coloring of the base map indicates relative elevations, based on data from the Mars Orbiter Laser Altimeter on NASA's Mars Global Surveyor. Reds and pinks are higher elevation (+3 km to +8 km); yellow is 0 km; greens and blues are lower elevation (down to −8 km). Whites (>+12 km) and browns (>+8 km) are the highest elevations. Axes are latitude and longitude; note poles are not shown.

Mars Exploration Rover.

Mars Exploration Rover B Opportunity

Opportunity is a robotic rover on the planet Mars, active since 2004. It is the remaining rover in NASA's ongoing Mars Exploration Rover Mission. Launched from Earth on July 7, 2003, it landed on the Martian Meridiani Planum on January 25, 2004 at 05:05 Ground UTC (about 13:15 local time), three weeks after its twin *Spirit* (MER-A) touched down on the other side of the planet. On July 28, 2014, NASA announced that *Opportunity*, after having traveled over 40 km (25 mi) on the planet Mars, has set a new "off-world" record as the rover having driven the greatest distance, surpassing the previous record held by the Soviet Union's Lunokhod 2 rover that had traveled 39 km (24 mi).

Mars Science Laboratory Rover "Curiosity"

On 26 November 2011, NASA's Mars Science Laboratory mission was successfully launched for Mars. The mission successfully landed the robotic "Curiosity" rover on the surface of Mars in August 2012, whereupon the rover is currently helping to determine whether Mars could ever have supported life, and search for evidence of past or present life on Mars.

Mars Science Laboratory.

Chang'e 3

Chang'e 3 is a Chinese Moon mission that includes a robotic lunar rover. Launched in 2013, it is China's first lunar rover, part of the second phase of the Chinese Lunar Exploration Program undertaken by China National Space Administration (CNSA).

Planned Rover Missions

Chandrayaan 2

The Chandrayaan-II mission is a joint venture between India and Russia, consisting of a lunar orbiter and a lunar lander. An opportunity was given to students to design this rover. 150 students gave their designs but only 6 were selected. They gave a demonstration in NRSA and are going to ISRO.The Russian designed rover weighs 50 kg, will have six wheels and will run on solar power. It will land near one of the poles and will operate for a year, roving up to 150 km at a maximum speed of 360 m/h. The proposed launch year is 2018.

ExoMars Rover

The European Space Agency (ESA) is currently designing and carrying out early prototyping and testing of the ExoMars rover which is scheduled for launch in 2018.

Mars 2020 Rover Mission

The Mars 2020 rover mission is a Mars planetary rovermission concept under study by NASA with a possible launch in 2020. It is intended to investigate an astrobiologically relevant ancient environment on Mars, investigate its surface geological processes and history, including the assessment of its past habitability and potential for preservation of biosignatures within accessible geological materials.

Future Lunar Missions

As of 2009, NASA had developed a series of plans for future moon missions which called for rovers that have a far longer range than the Apollo rovers.

Robot-assisted Surgery

Robotic surgery, computer-assisted surgery, and robotically-assisted surgery are terms for technological developments that use robotic systems to aid in surgical procedures. Robotically-assisted surgery was developed to overcome the limitations of pre-existing minimally-invasive surgical procedures and to enhance the capabilities of surgeons performing open surgery.

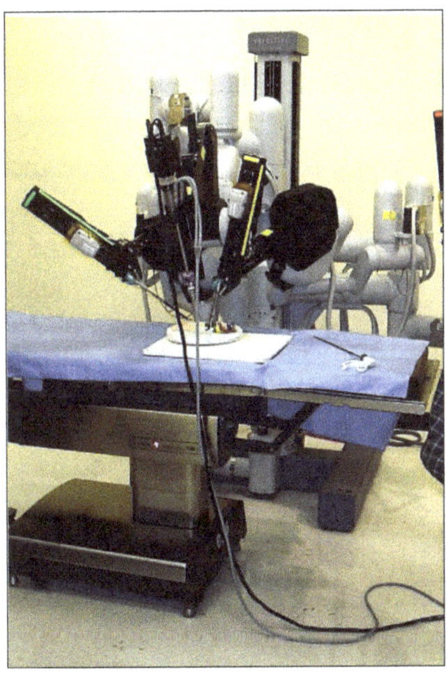

A robotically assisted surgical system used for prostatectomies, cardiac valve repair and gynecologic surgical procedures.

In the case of robotically-assisted minimally-invasive surgery, instead of directly moving the instruments, the surgeon uses one of two methods to control the instruments; either a direct telemanipulator or through computer control. A telemanipulator is a remote manipulator that allows the surgeon to perform the normal movements associated with the surgery whilst the robotic arms carry out those movements using end-effectors and manipulators to perform the actual surgery on the patient. In computer-controlled systems the surgeon uses a computer to control the robotic arms and its end-effectors, though these systems can also still use telemanipulators for their input. One advantage of using the computerised method is that the surgeon does not have to be present, but can be anywhere in the world, leading to the possibility for remote surgery.

In the case of enhanced open surgery, autonomous instruments (in familiar configurations) replace traditional steel tools, performing certain actions (such as rib spreading) with much smoother, feedback-controlled motions than could be achieved by a human hand. The main object of such smart instruments is to reduce or eliminate the tissue trauma traditionally associated with open surgery without requiring more than a few minutes' training on the part of surgeons. This approach seeks to improve open surgeries, particularly cardio-thoracic, that have so far not benefited from minimally-invasive techniques.

Robotic surgery has been criticized for its expense, by one estimate costing $1,500 to $2000 more per patient.

Comparison to Traditional Methods

Major advances aided by surgical robots have been remote surgery, minimally invasive surgery and unmanned surgery. Due to robotic use, the surgery is done with precision, miniaturization, smaller incisions; decreased blood loss, less pain, and quicker healing time. Articulation beyond normal manipulation and three-dimensional magnification helps resulting in improved ergonomics. Due to these techniques there is a reduced duration of hospital stays, blood loss, transfusions, and use of pain medication. The existing open surgery technique has many flaws like limited access to surgical area, long recovery time, long hours of operation, blood loss, surgical scars and marks.

The robot normally costs $1,390,000 and while its disposable supply cost is normally $1,500 per procedure, the cost of the procedure is higher. Additional surgical training is needed to operate the system. Numerous feasibility studies have been done to determine whether the purchase of such systems are worthwhile. As it stands, opinions differ dramatically. Surgeons report that, although the manufacturers of such systems provide training on this new technology, the learning phase is intensive and surgeons must operate on twelve to eighteen patients before they adapt. During the training phase, minimally invasive operations can take up to twice as long as traditional surgery, leading to operating room tie ups and surgical staffs keeping patients under anesthesia for longer periods. Patient surveys indicate they chose the procedure based on expectations of decreased morbidity, improved outcomes, reduced blood loss and less pain. Higher expectations may explain higher rates of dissatisfaction and regret.

Compared with other minimally invasive surgery approaches, robot-assisted surgery gives the surgeon better control over the surgical instruments and a better view of the surgical site. In addition, surgeons no longer have to stand throughout the surgery and do not tire as quickly. Naturally occurring hand tremors are filtered out by the robot's computer software. Finally, the surgical robot can continuously be used by rotating surgery teams.

Critics of the system, including the American Congress of Obstetricians and Gynecologists, say there is a steep learning curve for surgeons who adopt use of the system and that there's a lack of studies that indicate long-term results are superior to results following traditional laparoscopic surgery. Articles in the newly created *Journal of Robotic Surgery* tend to report on one surgeon's experience.

A Medicare study found that some procedures that have traditionally been performed with large incisions can be converted to "minimally invasive" endoscopic procedures with the use of the Da Vinci Surgical System, shortening length-of-stay in the hospital and reducing recovery times. But because of the hefty cost of the robotic system it is not clear that it is cost-effective for hospitals and physicians despite any benefits to patients since there is no additional reimbursement paid by the government or insurance companies when the system is used.

Robot-assisted pancreatectomies have been found to be associated with "longer operating time, lower estimated blood loss, a higher spleen-preservation rate, and shorter hospital stay[s]" than laparoscopic pancreatectomies; there was "no significant difference in transfusion, conversion to open surgery, overall complications, severe complications, pancreatic fistula, severe pancreatic fistula, ICU stay, total cost, and 30-day mortality between the two groups."

Uses

General Surgery

In early 2000 the field of general surgical interventions with the daVinci device was explored by surgeons at Ohio State University. Reports were published in esophageal and pancreatic surgery for the first time in the world and further data was subsequently published by Horgan and his group at the University of Illinois and then later at the same institution by others. In 2007, the University of Illinois at Chicago medical team, led by Prof. Pier Cristoforo Giulianotti, reported a pancreatectomy and also the Midwest's first fully robotic Whipple surgery. In April 2008, the same team of surgeons performed the world's first fully minimally invasive liver resection for living donor transplantation, removing 60% of the patient's liver, yet allowing him to leave the hospital just a couple of days after the procedure, in very good condition. Furthermore, the patient can also leave with less pain than a usual surgery due to the four puncture holes and not a scar by a surgeon.

Cardiothoracic Surgery

Robot-assisted MIDCAB and Endoscopic coronary artery bypass (TECAB) operations are being performed with the Da Vinci system. Mitral valve repairs and replacements have been performed. The Ohio State University, Columbus has performed CABG, mitral valve, esophagectomy, lung resection, tumor resections, among other robotic assisted procedures and serves as a training site for other surgeons. In 2002, surgeons at the Cleveland Clinic in Florida reported and published their preliminary experience with minimally invasive "hybrid" procedures. These procedures combined robotic revascularization and coronary stenting and further expanded the role of robots in coronary bypass to patients with disease in multiple vessels. Ongoing research on the outcomes of robotic assisted CABG and hybrid CABG is being done.

Cardiology and Electrophysiology

The Stereotaxis Magnetic Navigation System (MNS) has been developed to increase precision and safety in ablation procedures for arrhythmias and atrial fibrillation while reducing radiation exposure for the patient and physician, and the system utilizes two magnets to remotely steerable catheters. The system allows for automated 3-D mapping of the heart and vasculature, and MNS has also been used in interventional cardiology for guiding stents and leads in PCI and CTO procedures, proven to reduce contrast usage and access tortuous anatomy unreachable by manual navigation. Dr. Andrea Natale has referred to the new Stereotaxis procedures with the magnetic irrigated catheters as "revolutionary."

The Hansen Medical Sensei robotic catheter system uses a remotely operated system of pulleys to navigate a steerable sheath for catheter guidance. It allows precise and more forceful positioning of catheters used for 3-D mapping of the heart and vasculature. The system provides doctors with estimated force feedback information and feasible manipulation within the left atrium of the heart. The Sensei has been associated with mixed acute success rates compared to manual, commensurate with higher procedural complications, longer procedure times but lower fluoroscopy dosage to the patient.

At present, three types of heart surgery are being performed on a routine basis using robotic surgery systems. These three surgery types are:

- Atrial septal defect repair – the repair of a hole between the two upper chambers of the heart,
- Mitral valve repair – the repair of the valve that prevents blood from regurgitating back into the upper heart chambers during contractions of the heart,
- Coronary artery bypass – rerouting of blood supply by bypassing blocked arteries that provide blood to the heart.

As surgical experience and robotic technology develop, it is expected that the applications of robots in cardiovascular surgery will expand.

Colon and rectal surgery.

Many studies have been undertaken in order to examine the role of robotic procedures in the field of colorectal surgery.

Results to date indicate that robotic-assisted colorectal procedures outcomes are "no worse" than the results in the now "traditional" laparoscopic colorectal operations. Robotic-assisted colorectal surgery appears to be safe as well. Most of the procedures have been performed for malignant colon and rectal lesions. However, surgeons are now moving into resections for diverticulitis and non-resective rectopexies (attaching the colon to the sacrum in order to treat rectal prolapse.).

When evaluated for several variables, robotic-assisted procedures fare equally well when compared with laparoscopic, or open abdominal operations. Study parameters have looked at intraoperative patient preparation time, length of time to perform the operation, adequacy of the removed surgical specimen with respect to clear surgical margins and number of lymph nodes removed, blood loss, operative or postoperative complications and long-term results.

More difficult to evaluate are issues related to the view of the operative field, the types of procedures that should be performed using robotic assistance and the potential added cost for a robotic operation.

Many surgeons feel that the optics of the 3-dimensional, two camera stereo optic robotic system are superior to the optical system used in laparoscopic procedures. The pelvic nerves are clearly visualized during robotic-assisted procedures. Less clear however is whether or not these supposedly improved optics and visualization improve patient outcomes with respect to postoperative impotence or incontinence, and whether long-term patient survival is improved by using the 3-dimensional optic system. Additionally, there is often a need for a wider, or "larger" view of the operative field than is routinely provided during robotic operations., The close-up view of the area under dissection may hamper visualization of the "bigger view", especially with respect to ureteral protection.

Questions remain unanswered, even after many years of experience with robotic-assisted colorectal operations. Ongoing studies may help clarify many of the issues of confusion associated with this novel surgical approach.

Gastrointestinal Surgery

Multiple types of procedures have been performed with either the 'Zeus' or da Vinci robot systems, including bariatric surgery and gastrectomy for cancer. Surgeons at various universities initially published case series demonstrating different techniques and the feasibility of GI surgery using the robotic devices. Specific procedures have been more fully evaluated, specifically esophageal fundoplication for the treatment of gastroesophageal reflux and Heller myotomy for the treatment of achalasia.

Other gastrointestinal procedures including colon resection, pancreatectomy, esophagectomy and robotic approaches to pelvic disease have also been reported.

Gynecology

Robotic surgery in gynecology is of uncertain benefit with it being unclear if it affects rates of complications. Gynecologic procedures may take longer with robot-assisted surgery but may be associated with a shorter hospital stay following hysterectomy. In the United States, robotic-assisted hysterectomy for benign conditions has been shown to be more expensive than conventional laparoscopic hysterectomy, with no difference in overall rates of complications.

This includes the use of the da Vinci surgical system in benign gynecology and gynecologic oncology. Robotic surgery can be used to treat fibroids, abnormal periods, endometriosis, ovarian tumors, uterine prolapse, and female cancers. Using the robotic system, gynecologists can perform hysterectomies, myomectomies, and lymph node biopsies.

Neurosurgery

Several systems for stereotactic intervention are currently on the market. The NeuroMate was the first neurosurgical robot, commercially available in 1997. Originally developed in Grenoble by Alim-Louis_Benabid's team, it is now owned by Renishaw. With installations in the United States, Europe and Japan, the system has been used in 8000 stereotactic brain surgeries by 2009. IMRIS Inc.'s SYMBIS(TM) Surgical System will be the version of NeuroArm, the world's first MRI-compatible surgical robot, developed for world-wide commercialization. Medtech's Rosa is being used by several institutions, including the Cleveland Clinic in the U.S, and in Canada at Sherbrooke University and the Montreal Neurological Institute and Hospital in Montreal (MNI/H). Between June 2011 and September 2012, over 150 neurosurgical procedures at the MNI/H have been completed robotized stereotaxy, including in the placement of depth electrodes in the treatment of epilepsy, selective resections, and stereotaxic biopsies.

Ophthalmology

The first robotic operation inside the eye took place at the John Radcliffe Hospital in Oxford on 9 September 2016. The robot was developed by Preceyes BV and the surgery was performed by Robert MacLaren, Professor of Ophthalmology at the University of Oxford. They operated successfully on the Reverend Dr William Beaver, a 70 year old man with a macular hole. The researchers commented that robotic eye surgery has taken longer to develop due to the need to miniaturise the components. They observed that the robot could make controlled movements inside the eye as small as 100th of a millimetre.

Orthopedics

The ROBODOC system was released in 1992 by Integrated Surgical Systems, Inc. which merged into CUREXO Technology Corporation. Also, The Acrobot Company Ltd. developed the "Acrobot Sculptor", a robot that constrained a bone cutting tool to a pre-defined volume. The "Acrobot Sculptor" was sold to Stanmore Implants in August 2010. Stanmore received FDA clearance in February 2013 for US surgeries but sold the Sculptor to Mako Surgical in June 2013 to resolve a patent infringement lawsuit. Another example is the CASPAR robot produced by U.R.S.which is used for total hip replacement, total knee replacement and anterior cruciate ligament reconstruction. MAKO Surgical Corp (founded 2004) produces the RIO (Robotic Arm Interactive Orthopedic System) which combines robotics, navigation, and haptics for both partial knee and total hip replacement surgery. Blue Belt Technologies received FDA clearance in November 2012 for the Navio™ Surgical System. The Navio System is a navigated, robotics-assisted surgical system that uses a CT free approach to assist in partial knee replacement surgery.

Pediatrics

Surgical robotics has been used in many types of pediatric surgical procedures including: tracheoesophageal fistula repair, cholecystectomy, nissen fundoplication, morgagni's hernia repair, kasai portoenterostomy, congenital diaphragmatic hernia repair, and others. On 17 January 2002, surgeons at Children's Hospital of Michigan in Detroit performed the nation's first advanced computer-assisted robot-enhanced surgical procedure at a children's hospital.

The Center for Robotic Surgery at Children's Hospital Boston provides a high level of expertise in pediatric robotic surgery. Specially-trained surgeons use a high-tech robot to perform complex and delicate operations through very small surgical openings. The results are less pain, faster recoveries, shorter hospital stays, smaller scars, and happier patients and families.

In 2001, Children's Hospital Boston was the first pediatric hospital to acquire a surgical robot. Today, surgeons use the technology for many procedures and perform more pediatric robotic operations than any other hospital in the world. Children's Hospital physicians have developed a number of new applications to expand the use of the robot, and train surgeons from around the world on its use.

Radiosurgery

The CyberKnife Robotic Radiosurgery System uses image guidance and computer controlled robotics to treat tumors throughout the body by delivering multiple beams of high-energy radiation to the tumor from virtually any direction. The system uses a German KUKA KR 240. Mounted on the robot is a compact X-band linac that produces 6MV X-ray radiation. Mounting the radiation source on the robot allows very fast repositioning of the source, which enables the system to deliver radiation from many different directions without the need to move both the patient and source as required by current gantry configurations.

Transplant Surgery

Transplant surgery (organ transplantation) has been considered as highly technically demanding and virtually unobtainable by means of conventional laparoscopy. For many years, transplant patients were unable to benefit from the advantages of minimally invasive surgery. The development of robotic technology and its associated high resolution capabilities, three dimensional visual system, wrist type motion and fine instruments, gave opportunity for highly complex procedures to be completed in a minimally invasive fashion. Subsequently, the first fully robotic kidney transplantations were performed in the late 2000s. After the procedure was proven to be feasible and safe, the main emerging challenge was to determine which patients would benefit most from this robotic technique. As a result, recognition of the increasing prevalence of obesity amongst patients with kidney failure on hemodialysis posed a significant problem. Due to the abundantly higher risk of complications after traditional open kidney transplantation, obese patients were frequently denied access to transplantation, which is the premium treatment for end stage kidney disease. The use of the robotic-assisted approach has allowed kidneys to be transplanted with minimal incisions, which has virtually alleviated wound complications and significantly shortened the recovery period. The University of Illinois Medical Center reported the largest series of 104 robotic-assisted kidney transplants for obese recipients (mean body mass index > 42). Amongst this group of patients, no wound infections were observed and the function of transplanted kidneys was excellent. In this way, robotic kidney transplantation could be considered as the biggest advance in surgical technique for this procedure since its creation more than half a century ago.

Urology

Robotic surgery in the field of urology has become very popular, especially in the United States. It has been most extensively applied for excision of prostate cancer because of difficult anatomical access. It is also utilized for kidney cancer surgeries and to lesser extent surgeries of the bladder.

As of 2014, there is little evidence of increased benefits compared to standard surgery to justify the increased costs. Some have found tentative evidence of more complete removal of cancer and less side effects from surgery for prostatectomy.

In 2000, the first robot-assisted laparoscopic radical prostatectomy was performed.

Vascular Surgery

In September 2010, the first robotic operations with Hansen Medical's Magellan Robotic System at the femoral vasculature were performed at the University Medical Centre Ljubljana (UMC Ljubljana), Slovenia. The research was led by Borut Geršak, the head of the Department of Cardiovascular Surgery at the centre. Geršak explained that the robot used was the first true robot in the history of robotic surgery, meaning the user interface was not resembling surgical instruments and the robot was not simply imitating the movement of human hands but was guided by pressing buttons, just like one would play a video game. The robot was imported to Slovenia from the United States.

Miniature Robotics

As scientists seek to improve the versatility and utility of robotics in surgery, some are attempting to miniaturize the robots. For example, the University of Nebraska Medical Center has led a multi-campus effort to provide collaborative research on mini-robotics among surgeons, engineers and computer scientists.

History

The first robot to assist in surgery was the *Arthrobot*, which was developed and used for the first time in Vancouver in 1983. Intimately involved were biomedical engineer, Dr. James McEwen, Geof Auchinleck, a UBC engineering physics grad, and Dr. Brian Day as well as a team of engineering students. The robot was used in an orthopaedic surgical procedure on 12 March 1984, at the UBC Hospital in Vancouver. Over 60 arthroscopic surgical procedures were performed in the first 12 months, and a 1985 National Geographic video on industrial robots, *The Robotics Revolution*, featured the device. Other related robotic devices developed at the same time included a surgical scrub nurse robot, which handed operative instruments on voice command, and a medical laboratory robotic arm. A YouTube video entitled *Arthrobot* illustrates some of these in operation.

In 1985 a robot, the Unimation Puma 200, was used to place a needle for a brain biopsy using CT guidance. In 1992, the PROBOT, developed at Imperial College London, was used to perform prostatic surgery by Dr. Senthil Nathan at Guy's and St Thomas' Hospital, London. This was the first pure robotic surgery in the world. Also the Robot Puma 560, a robot developed in 1985 by Kwoh et al. Puma 560 was used to perform neurosurgical biopsies with greater precision. Just like with any other technological innovation, this system led to the development of new and improved surgical robot called PROBOT. The PROBOT was specifically designed for transurethral resection of the prostate. Meanwhile, when PROBOT was being developed, ROBODOC, a robotic system designed to assist hip replacement surgeries was the first surgical robot that was approved by the FDA. The ROBODOC from Integrated Surgical Systems (working closely with IBM) was introduced in 1992 to mill out precise fittings in the femur for hip replacement. The purpose of the ROBODOC was to replace the previous method of carving out a femur for an implant, the use of a mallet and broach/rasp.

Further development of robotic systems was carried out by SRI International and Intuitive Surgical with the introduction of the da Vinci Surgical System and Computer Motion with the *AESOP* and the ZEUS robotic surgical system. The first robotic surgery took place at The Ohio State University Medical Center in Columbus, Ohio under the direction of Robert E. Michler. Examples of using ZEUS include a fallopian tube reconnection in July 1998, a *beating heart* coronary artery bypass graft in October 1999, and the Lindbergh Operation, which was a cholecystectomy performed remotely in September 2001.

The original telesurgery robotic system that the da Vinci was based on was developed at SRI International in Menlo Park with grant support from DARPA and NASA. Although the telesurgical robot was originally intended to facilitate remotely performed surgery in battlefield and other remote environments, it turned out to be more useful for minimally invasive on-site surgery. The patents for the early prototype were sold to Intuitive Surgical in Mountain View, California. The da Vinci senses the surgeon's hand movements and translates them electronically into scaled-down micro-movements to manipulate the tiny proprietary instruments. It also detects and filters out any tremors in the surgeon's hand movements, so that they are not duplicated robotically. The camera used in the system provides a true stereoscopic picture transmitted to a surgeon's console. Examples of using the da Vinci system include the first robotically assisted heart bypass (performed in Germany) in May 1998, and the first performed in the United States in September 1999; and the first all-robotic-assisted kidney transplant, performed in January 2009. The da Vinci Si was released in April 2009, and initially sold for $1.75 million.

In May 2006 the first artificial intelligence doctor-conducted unassisted robotic surgery on a 34-year-old male to correct heart arythmia. The results were rated as better than an above-average human surgeon. The machine had a database of 10,000 similar operations, and so, in the words of its designers, was "more than qualified to operate on any patient". In August 2007, Dr. Sijo Parekattil of the Robotics Institute and Center for Urology (Winter Haven Hospital and University of Florida) performed the first robotic assisted microsurgery procedure denervation of the spermatic cord for chronic testicular pain. In February 2008, Dr. Mohan S. Gundeti of the University of Chicago Comer Children's Hospital performed the first robotic pediatric neurogenic bladder reconstruction.

On 12 May 2008, the first image-guided MR-compatible robotic neurosurgical procedure was performed at University of Calgary by Dr. Garnette Sutherland using the NeuroArm. In June 2008, the German Aerospace Centre (DLR) presented a robotic system for minimally invasive surgery, the MiroSurge. In September 2010, the Eindhoven University of Technology announced the development of the Sofie surgical system, the first surgical robot to employ force feedback. In September 2010, the first robotic operation at the femoral vasculature was performed at the University Medical Centre Ljubljana by a team led by Borut Geršak.

References

- Denaro, Bob (1 April 2016). "ITS International" (PDF). Civil Maps - Automated Vehicle: Myth vs. Reality. ITS International. Retrieved 22 June 2016
- Zhang, Benjamin (2016-01-10). "ELON MUSK: In 2 years your Tesla will be able to drive from New York to LA and find you". Automotive News. Retrieved 2016-01-12

- Dudley, David (January 2015). "The Driverless Car Is (Almost) Here; The self-driving car — a godsend for older Americans — is now on the horizon". AARP The Magazine. AARP. Retrieved 30 November 2015
- Adhikari, Richard (11 February 2016). "Feds Put AI in the Driver's Seat". Technewsworld. Retrieved 12 February 2016
- Nichols, Greg (13 February 2016). "NHTSA chief takes conservative view on autonomous vehicles: "If you had perfect, connected autonomous vehicles on the road tomorrow, it would still take 20 to 30 years to turn over the fleet."". ZDNet. Retrieved 17 February 2016
- Burn-Callander, Rebecca (11 February 2015). "This is the Lutz pod, the UK's first driverless car". Daily Telegraph. Retrieved 11 February 2015
- Miller, John (19 August 2014). "Self-Driving Car Technology's Benefits, Potential Risks, and Solutions". theenergycollective.com. Retrieved 4 June 2015
- Charlton, Alistair (2016-06-13). "Tesla Autopilot is 'trying to kill me', says Volvo R&D chief". International Business Times. Retrieved 2016-07-01
- Golson, Jordan (2016-04-27). "Volvo autonomous car engineer calls Tesla's Autopilot a 'wannabe'". The Verge. Retrieved 2016-07-01
- Bryner, Jeanna (12 January 2016). "Ford Takes Autonomous Cars for Snowy Test Drive". LiveScience. Retrieved 14 January 2016
- Abuelsamid, Sam (2016-07-01). "Tesla Autopilot Fatality Shows Why Lidar And V2V Will Be Necessary For Autonomous Cars". Forbes. Retrieved 2016-07-01
- Yadron, Danny; Tynan, Dan (2016-07-01). "Tesla driver dies in first fatal crash while using autopilot mode". The Guardian. San Francisco. Retrieved 2016-07-01
- Morris, David Paul (2016-07-01). "Highway patrol found DVD player in wreckage of fatal Tesla accident". Associated Press. CNBC. Retrieved 2016-07-01
- Office of Defects Investigations, NHTSA (2016-06-28). "ODI Resume - Investigation: PE 16-007" (PDF). National Highway Traffic Safety Administration (NHTSA). Retrieved 2016-07-02
- Chipman, Ian (2016-02-08). "Meet "Hedgehog": Engineers build cube-like rover for exploration of asteroids, comets". Phys.org. Retrieved 2016-02-11

7

Diverse Aspects of Robotics

Degrees of freedom of a mechanical system refers to the number of independent variables required to define the position of a rigid body in space. The ethical problems that occur with robots fall under roboethics. All these diverse principles and laws of robotics have been carefully analyzed in this chapter.

Degrees of Freedom

In physics, the degree of freedom (DOF) of a mechanical system is the number of independent parameters that define its configuration. It is the number of parameters that determine the state of a physical system and is important to the analysis of systems of bodies in mechanical engineering, aeronautical engineering, robotics, and structural engineering.

The position of a single car (engine) moving along a track has one degree of freedom because the position of the car is defined by the distance along the track. A train of rigid cars connected by hinges to an engine still has only one degree of freedom because the positions of the cars behind the engine are constrained by the shape of the track.

An automobile with highly stiff suspension can be considered to be a rigid body traveling on a plane (a flat, two-dimensional space). This body has three independent degrees of freedom consisting of two components of translation and one angle of rotation. Skidding or drifting is a good example of an automobile's three independent degrees of freedom.

The position and orientation of a rigid body in space is defined by three components of translation and three components of rotation, which means that it has six degrees of freedom.

The exact constraint mechanical design method manages the degrees of freedom to neither underconstrain nor overconstrain a device.

Motions and Dimensions

The position of an n-dimensional rigid body is defined by the rigid transformation, $[T] = [A, d]$, where d is an n-dimensional translation and A is an $n \times n$ rotation matrix, which has n translational degrees of freedom and $n(n - 1)/2$ rotational degrees of freedom. The number of rotational degrees of freedom comes from the dimension of the rotation group SO(n).

A non-rigid or deformable body may be thought of as a collection of many minute particles (infinite number of DOFs), this is often approximated by a finite DOF system. When motion involving large displacements is the main objective of study (e.g. for analyzing the motion of satellites), a deformable body may be approximated as a rigid body (or even a particle) in order to simplify the analysis.

The degree of freedom of a system can be viewed as the minimum number of coordinates required to specify a configuration. Applying this definition, we have:

1. For a single particle in a plane two coordinates define its location so it has two degrees of freedom.
2. A single particle in space requires three coordinates so it has three degrees of freedom.
3. Two particles in space have a combined six degrees of freedom.
4. If two particles in space are constrained to maintain a constant distance from each other, such as in the case of a diatomic molecule, then the six coordinates must satisfy a single constraint equation defined by the distance formula. This reduces the degree of freedom of the system to five, because the distance formula can be used to solve for the remaining coordinate once the other five are specified.

Six Degrees of Freedom

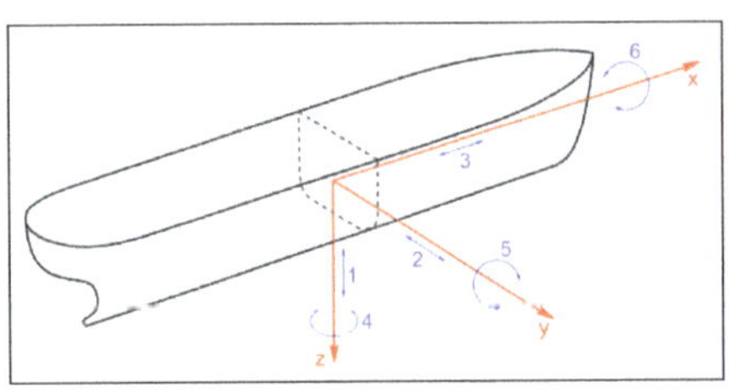

The six degrees of freedom of movement of a ship.

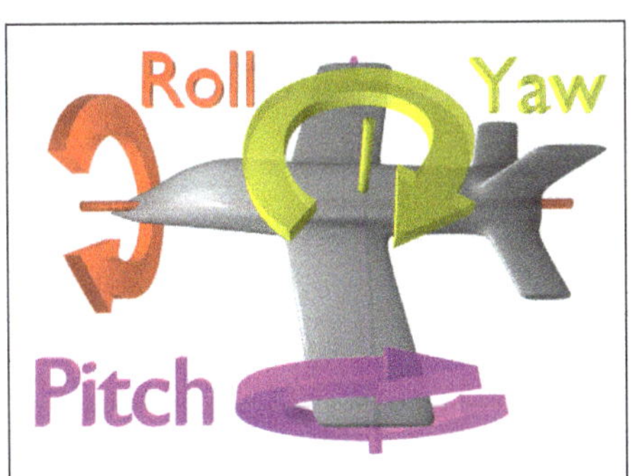

Attitude degrees of freedom for an airplane.

The motion of a ship at sea has the six degrees of freedom of a rigid body, and is described as:

Translation and Rotation:

1. Moving up and down (elevating/heaving).
2. Moving left and right (strafing/swaying).
3. Moving forward and backward (walking/surging).
4. Swivels left and right (yawing).
5. Tilts forward and backward (pitching).
6. Pivots side to side (rolling).

The trajectory of an airplane in flight has three degrees of freedom and its attitude along the trajectory has three degrees of freedom, for a total of six degrees of freedom.

Mobility Formula

The mobility formula counts the number of parameters that define the configuration of a set of rigid bodies that are constrained by joints connecting these bodies.

Consider a system of n rigid bodies moving in space has $6n$ degrees of freedom measured relative to a fixed frame. In order to count the degrees of freedom of this system, include the ground frame in the count of bodies, so that mobility is independent of the choice of the body that forms the fixed frame. Then the degree-of-freedom of the unconstrained system of $N = n + 1$ is:

$$M = 6n = 6(N-1),$$

because the fixed body has zero degrees of freedom relative to itself.

Joints that connect bodies in this system remove degrees of freedom and reduce mobility. Specifically, hinges and sliders each impose five constraints and therefore remove five degrees of freedom. It is convenient to define the number of constraints c that a joint imposes in terms of the joint's freedom f, where $c = 6 - f$. In the case of a hinge or slider, which are one degree of freedom joints, have $f = 1$ and therefore $c = 6 - 1 = 5$.

The result is that the mobility of a system formed from n moving links and j joints each with freedom f_i, $i = 1, \ldots, j$, is given by:

$$M = 6n - \sum_{i=1}^{j} (6 - f_i) = 6(N - 1 - j) + \sum_{i=1}^{j} f_i$$

Recall that N includes the fixed link.

There are two important special cases: (i) a simple open chain, and (ii) a simple closed chain. A single open chain consists of n moving links connected end to end by n joints, with one end connected to a ground link. Thus, in this case $N = j + 1$ and the mobility of the chain is:

$$M = \sum_{i=1}^{j} f_i$$

For a simple closed chain, n moving links are connected end-to-end by $n + 1$ joints such that the two ends are connected to the ground link forming a loop. In this case, we have $N = j$ and the mobility of the chain is:

$$M = \sum_{i=1}^{j} f_i - 6$$

An example of a simple open chain is a serial robot manipulator. These robotic systems are constructed from a series of links connected by six one degree-of-freedom revolute or prismatic joints, so the system has six degrees of freedom.

An example of a simple closed chain is the RSSR spatial four-bar linkage. The sum of the freedom of these joints is eight, so the mobility of the linkage is two, where one of the degrees of freedom is the rotation of the coupler around the line joining the two S joints.

Planar and Spherical Movement

It is common practice to design the linkage system so that the movement of all of the bodies are constrained to lie on parallel planes, to form what is known as a *planar linkage*. It is also possible to construct the linkage system so that all of the bodies move on concentric spheres, forming a *spherical linkage*. In both cases, the degrees of freedom of the links in each system is now three rather than six, and the constraints imposed by joints are now $c = 3 - f$.

In this case, the mobility formula is given by:

$$M = 3(N - 1 - j) + \sum_{i=1}^{j} f_i,$$

and the special cases become:

- Planar or spherical simple open chain,

$$M = \sum_{i=1}^{j} f_i,$$

- Planar or spherical simple closed chain,

$$M = \sum_{i=1}^{j} f_i - 3$$

An example of a planar simple closed chain is the planar four-bar linkage, which is a four-bar loop with four one degree-of-freedom joints and therefore has mobility $M = 1$.

Systems of Bodies

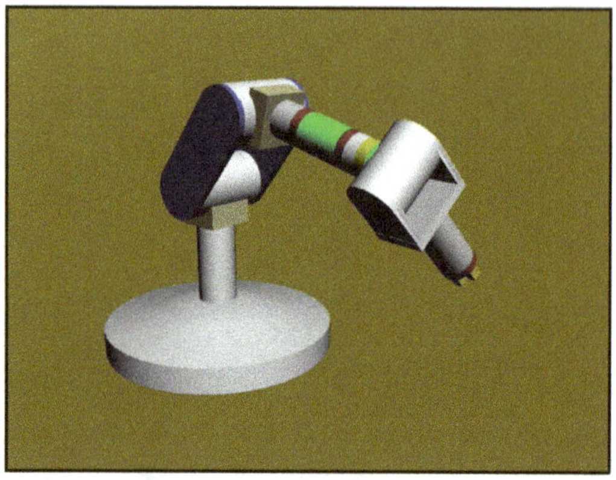

An articulated robot with six DOF in a kinematic chain.

A system with several bodies would have a combined DOF that is the sum of the DOFs of the bodies, less the internal constraints they may have on relative motion. A mechanism or linkage containing a number of connected rigid bodies may have more than the degrees of freedom for a single rigid body. Here the term *degrees of freedom* is used to describe the number of parameters needed to specify the spatial pose of a linkage.

A specific type of linkage is the open kinematic chain, where a set of rigid links are connected at joints; a joint may provide one DOF (hinge/sliding), or two (cylindrical). Such chains occur commonly in robotics, biomechanics, and for satellites and other space structures. A human arm is considered to have seven DOFs. A shoulder gives pitch, yaw, and roll, an elbow allows for pitch, and a wrist allows for pitch,yaw and roll . Only 3 of those movements would be necessary to move the hand to any point in space, but people would lack the ability to grasp things from different angles or directions. A robot (or object) that has mechanisms to control all 6 physical DOF is said to be holonomic. An object with fewer controllable DOFs than total DOFs is said to be non-holonomic, and an object with more controllable DOFs than total DOFs (such as the human arm) is said to be redundant. Although keep in mind that it is not redundant in the human arm because the two DOFs; wrist and shoulder, that represent the same movement; roll, supply each other since they can't do a full 360. The degree of freedom are like different movements that can be made.

In mobile robotics, a car-like robot can reach any position and orientation in 2-D space, so it needs 3 DOFs to describe its pose, but at any point, you can move it only by a forward motion and a steering angle. So it has two control DOFs and three representational DOFs; i.e. it is non-holonomic. A fixed-wing aircraft, with 3–4 control DOFs (forward motion, roll, pitch, and to a limited extent, yaw) in a 3-D space, is also non-holonomic, as it cannot move directly up/down or left/right.

A summary of formulas and methods for computing the degrees-of-freedom in mechanical systems has been given by Pennestri, Cavacece, and Vita.

Electrical Engineering

In electrical engineering *degrees of freedom* is often used to describe the number of directions in which a phased array antenna can form either beams or nulls. It is equal to one less than the number of elements contained in the array, as one element is used as a reference against which either constructive or destructive interference may be applied using each of the remaining antenna elements. Radar practice and communication link practice, with beam steering being more prevalent for radar applications and null steering being more prevalent for interference suppression in communication links.

Roboethics

Roboethics is a short expression for "ethics of robotics" or "robot ethics". It concerns ethical problems that occur with robots, such as whether robots pose a threat to humans in the long or short run, whether some uses of robots are problematic (such as in healthcare or as 'killer robots' in war), and how robots should be designed such as they act 'ethically' (this last concern is also called machine ethics). Robot ethics is a sub-field of ethics of technology, specifically information technology, and it has close links to legal as well as socio-economic concerns.

While the issues are as old as the word *robot*, serious academic discussions started around the year 2000, e.g. in 2002, an Atelier funded by the European Robotics Research Network set up a road map effectively divided ethics of artificial intelligence into two sub-fields (ethics for and ethics with robots). The "First International Symposium on Roboethics" was held in (Sanremo, Italy, 2004).

Roboethics requires the combined commitment of experts of several disciplines, who, working in transnational projects, committees, commissions, have to adjust laws and regulations to the problems resulting from the scientific and technological achievements in Robotics and AI. The main fields involved in roboethics are: robotics, computer science, artificial intelligence, philosophy, ethics, theology, biology, physiology, cognitive science, neurosciences, law, sociology, psychology, and industrial design.

History & Events

Since antiquity, the discussion of ethics in relation to the treatment of non-human and even non-living things and their potential "spirituality" have been discussed. With the development of machinery and eventually robots, this philosophy was also applied to robotics. The first publication directly addressing roboethics was developed by Isaac

Asimov as his Three Laws of Robotics in 1942, in the context of his science fiction works, although the term “roboethics” was probably coined by Gianmarco Veruggio.

The Roboethic guidelines were developed during some important robotics events and projects:

- 1942, Asimov’s short story “Runaround” explicitly states his Three Laws for the first time. Those “Laws” get reused for later works of robot-related science fiction by Asimov.
- 2004, First International Symposium on Roboethics, 30–31 January 2004, Villa Nobel, Sanremo, Italy, organized by School of Robotics, where, the word Roboethics was officially used for the first time.
- 2004, IEEE-RAS established a Technical Committee on Roboethics.
- 2004, Fukuoka World Robot Declaration, issued on February 25, 2004 from Fukuoka, Japan.
- 2005, ICRA05 (International Conference on Robotics and Automation), Barcelona: the IEEE RAS TC on Roboethics organized a Workshop on Roboethics.
- 2005–2006, E.C. Euron Roboethics Atelier (Genoa, Italy, February/March 2006). The Euron Project, coordinated by School of Robotics, involved a large number of roboticists and scholars of humanities who produced the first Roadmap for a Roboethics.
- 2006, BioRob2006 (The first IEEE / RAS-EMBS International Conference on Biomedical Robotics and Bio-mechatronics), Pisa, Italy, February 20, 2006: Mini symposium on Roboethics.
- 2006, International Workshop “Ethics of Human Interaction with Robotic, Bionic, and AI Systems: Concepts and Policies”, Naples, 17–18 October 2006. The workshop was supported by the ETHICBOTS European Project.
- 2007 ICRA07 (International Conference on Robotics and Automation), Rome: the IEEE RAS TC on Roboethics organized a Workshop on Roboethics.
- 2007 ICAIL’07,International Conference on Artificial Intelligence and Law, Stanford University, Palo Alto, USA, 4–8 June.
- 2007 International European Conference on Computing and Philosophy E-CAP ‘07, University of Twente, Netherlands, 21–23 June 2007. Track “Roboethics”.
- 2007 Computer Ethics Philosophical Enquiry CEPE ‘07, University of San Diego, USA,12–14 July 2007. Topic “Roboethics”.
- 2008 INTERNATIONAL SYMPOSIUM ROBOTICS: NEW SCIENCE, Thursday FEBRUARY 20th, 2008, Via della Lungara 10 – ROME - ITALY.

- 2009 ICRA09 (International Conference on Robotics and Automation), Kobe, Japan: the IEEE RAS TC on Roboethics organized a Workshop on Roboethics.
- 2012 We Robot 2012, University of Miami, FL, USA .
- 2013 Workshop on Robot Ethics, University of Sheffield, Feb 2013.
- 2013 We Robot 2013 - Getting Down to Business, Stanford University.
- 2014 We Robot 2014 - Risks and Opportunities, University of Miami, FL, USA.
- 2016 Ethical and Moral Considerations in Non-Human Agents, Stanford Spring Symposium, AAAI Association for the Advancement of Artificial Intelligence.

Events in the field are announced e.g. by the euRobotics ELS topics group, and by RoboHub.

In Popular Culture

Roboethics as a science or philosophical topic has not made any strong cultural impact, but is a common theme in science fiction literature and films. One of the most popular films depicting the potential misuse of robotic and AI technology is *The Matrix*, depicting a future where the lack of roboethics brought about the destruction of the human race. An animated film based on *The Matrix*, the *Animatrix*, focused heavily on the potential ethical issues between humans and robots. Many of the Animatrix's animated shorts are also named after Isaac Asimov's fictional stories.

Although not a part of roboethics *per se*, the ethical behavior of robots themselves has also been a joining issue in roboethics in popular culture. The *Terminator* series focuses on robots run by an uncontrolled AI program with no restraint on the termination of its enemies. This series too has the same futuristic plot as *The Matrix* series, where robots have taken control. The most famous case of robots or computers without programmed ethics is HAL 9000 in the *Space Odyssey* series, where HAL (a computer with advance AI capabilities who monitors and assists humans on a space station) kills all the humans on board to ensure the success of the assigned mission after his own life is threatened.

Humanoid

A humanoid (from English *human* and *-oid* "resembling") is something that has an appearance resembling a human being. The earliest recorded use of the term, in 1870, referred to indigenous peoples in areas colonized by Europeans. By the 20th century, the term came to describe fossils which were morphologically similar, but not identical, to those of the human skeleton. Although this usage was common in the sciences for much of the 20th century, it is now considered rare. More generally, the term can refer to anything with uniquely human characteristics and/or adaptations, such as possessing opposable anterior forelimb-appendages (thumbs), visible spectrum-binocular vision (having two eyes), or biomechanic digitigrade-bipedalism (the ability to walk on heels in an upright position).

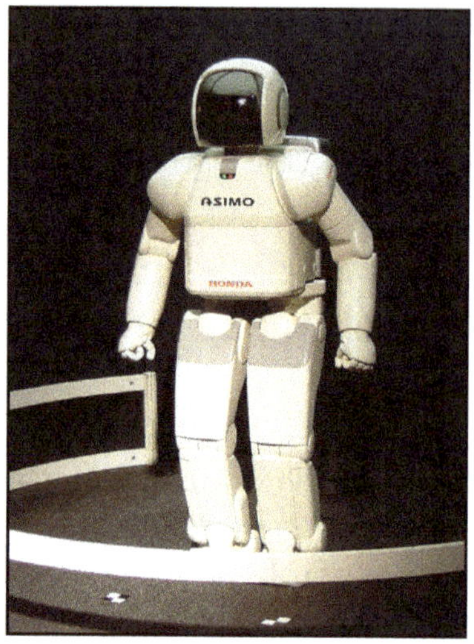

Honda's ASIMO is an example of a humanoid robot.

In Theoretical Convergent Evolution

Although there are no known humanoid species outside the genus *Homo*, the theory of convergent evolution speculates that different species may evolve similar traits, and in the case of a humanoid these traits may include intelligence and bipedalism and other humanoid skeletal changes, as a result of similar evolutionary pressures. American psychologist and Dinosaur intelligence theorist Harry Jerison suggested the possibility of sapient dinosaurs. In a 1978 presentation at the American Psychological Association, he speculated that dromiceiomimus could have evolved into a highly intelligent species like human beings. In his book, *Wonderful Life*, Stephen Jay Gould argues that if the tape of life were re-wound and played back, life would have taken a very different course. Simon Conway Morris counters this argument, arguing that convergence is a dominant force in evolution and that since the same environmental and physical constraints act on all life, there is an "optimum" body plan that life will inevitably evolve toward, with evolution bound to stumble upon intelligence, a trait of primates, crows, and dolphins, at some point.

In 1982, Dale Russell, curator of vertebrate fossils at the National Museum of Canada in Ottawa, conjectured a possible evolutionary path that might have been taken by the dinosaur *Troodon* had it not perished in the Cretaceous–Paleogene extinction event 66 million years ago, suggesting that it could have evolved into intelligent beings similar in body plan to humans, becoming a *humanoid* of dinosaur origin. Over geologic time, Russell noted that there had been a steady increase in the encephalization quotient or EQ (the relative brain weight when compared to other species with the same body weight) among the dinosaurs. Russell had discovered the first Troodontid skull, and noted that, while its EQ was low compared to humans, it was six times higher than that of other dinosaurs. If the trend in *Troodon* evolution had continued to the present, its brain case could by now measure 1,100 cm^3; comparable to that of a human. Troodontids had semi-manipulative fingers, able to grasp and hold objects to a certain degree, and binocular vision.

A model of the hypothetical Dinosauroid, Dinosaur Museum, Dorchester

Russell proposed that this "Dinosauroid", like most dinosaurs of the troodontid family, would have had large eyes and three fingers on each hand, one of which would have been partially opposed. As with most modern reptiles (and birds), he conceived of its genitalia as internal. Russell speculated that it would have required a navel, as a placenta aids the development of a large brain case. However, it would not have possessed mammary glands, and would have fed its young, as birds do, on regurgitated food. He speculated that its language would have sounded somewhat like bird song.

Russell's thought experiment has been met with criticism from other paleontologists since the 1980s, many of whom point out that his Dinosauroid is overly anthropomorphic. Gregory S. Paul (1988) and Thomas R. Holtz, Jr., consider it "suspiciously human" (Paul, 1988) and Darren Naish has argued that a large-brained, highly intelligent troodontid would retain a more standard theropod body plan, with a horizontal posture and long tail, and would probably manipulate objects with the snout and feet in the manner of a bird, rather than with human-like "hands".

In Robotics

A *humanoid robot* is a robot that is based on the general structure of a human, such as a robot that walks on two legs and has an upper torso, or a robot that has two arms, two legs and a head. A humanoid robot does not necessarily look convincingly like a real person, for example the ASIMO humanoid robot has a helmet instead of a face.

An *android* (male) or *gynoid* (female) is a humanoid robot designed to look as much like a real person as possible, although these words are frequently perceived to be synonymous with humanoid.

While there are many humanoid robots in fictional stories, some real humanoid robots have been developed since the 1990s, and some real human-looking android robots have been developed since 2002.

Similarly to robots, virtual avatars may also be called humanoid when resembling humans.

In Mythology

Deities are often imagined in human shape (also known as "anthropotheism"), sometimes as hybrids (especially the gods of Ancient Egyptian religion). A fragment by the Greek poet Xenophanes describes this tendency,

...Men make gods in their own image; those of the Ethiopians are black and narrow-nosed, those of the Thracians have blue eyes and red hair.

In animism in general, the spirits innate in certain objects (like the Greek nymphs) are typically depicted in human shape, e.g. spirits of trees (Dryads), of the woodlands (the hybrid fauns), of wells or waterways (Nereids, Necks), etc.

In Science Fiction

With regard to extraterrestrials in fiction, the term humanoid is most commonly used to refer to alien beings with a body plan that is generally like that of a human, including upright stance and bipedalism.

Many aliens in television and science fiction films are presented as humanoid. This is usually attributed to budget constraints, as human actors can more easily portray human-like aliens.

Fictional Extraterrestrials

In much of science fiction, the reason for the abundance of humanoid aliens is not explained and requires suspension of disbelief. In some cases, however, explanations have been offered for this. In *Star Trek*, the abundance of humanoid aliens within the *Star Trek* universe is explained by advancing the story of a primordial humanoid civilization, the Ancient humanoids, who seeded the galaxy with genetically-engineered cells to guide the evolution of life on a multitude of worlds toward a humanoid form. In the television series *Stargate SG-1*, the Jaffa are explained as being an hundred-thousand year offshoot of humanity bred by the Goa'uld to suit their purposes, hence their almost-human appearance and physiology, while many other "alien" characters are actually the descendants of human-slaves who were removed from Earth by the Goa'uld. Here on Earth, any species segregated from the main genus for at least 10k years may be considered a new sub-species; any humans isolated on multiple planets after 100k+ years of adaptations would most certainly seem "alien" to Earthlings. Similarly, in its spin-off show *Stargate Atlantis*, the explanation offered for the humanoid appearance of the Wraith is that the Wraith evolved from a parasite which incorporated human DNA into its own genome after feeding on humans, giving the Wraith their present form.

In Ufology

In the field of ufology, *humanoid* refers to any of the claimed extraterrestrials which abduct human victims, such as the Greys, the Reptilians, Nordics, and Martians.

In Fantasy

In fantasy settings the term *humanoid* is used to refer to a human-like fantastical creature, such as a dwarf, elf, gnome, halfling, goblin, troll, orc or an ogre, and Bigfoot . In some cases, such as older versions of the game Dungeons and Dragons, a distinction is made between demi-humans, which are relatively similar to humans, and humanoids, which exhibit larger differences from humans. Animals that are humanoid are also shown in fantasy. Humanoids are also used in some old horror movies, for example in Creature From the Black Lagoon, made in 1954 by Jack Arnold.

Cyborg

A cyborg (short for "cybernetic organism") is a being with both organic and biomechatronic body parts. The term was coined in 1960 by Manfred Clynes and Nathan S. Kline.

The term cyborg is not the same thing as bionic, biorobot or android; it applies to an organism that has restored function or enhanced abilities due to the integration of some artificial component or technology that relies on some sort of feedback. While cyborgs are commonly thought of as mammals, including humans, they might also conceivably be any kind of organism.

D. S. Halacy's *Cyborg: Evolution of the Superman* in 1965 featured an introduction which spoke of a "new frontier" that was "not merely space, but more profoundly the relationship between 'inner space' to 'outer space' – a bridge...between mind and matter."

In popular culture, some cyborgs may be represented as visibly mechanical (e.g., the Cybermen in the *Doctor Who* franchise or The Borg from *Star Trek* or Darth Vader from *Star Wars*) or as almost indistinguishable from humans (e.g., the "Human" Cylons from the re-imagining of *Battlestar Galactica*, etc.). Cyborgs in fiction often play up a human contempt for over-dependence on technology, particularly when used for war, and when used in ways that seem to threaten free will. Cyborgs are also often portrayed with physical or mental abilities far exceeding a human counterpart (military forms may have inbuilt weapons, among other things).

Overview

According to some definitions of the term, the physical attachments humanity has with even the most basic technologies have already made them cyborgs. In a typical example, a human with an artificial cardiac pacemaker or implantable cardioverter-defibrillator would be considered a cyborg, since these devices measure voltage potentials in the body, perform signal processing, and can deliver electrical stimuli, using this synthetic feedback mechanism to keep that person alive. Implants, especially cochlear implants, that combine mechanical modification with any kind of feedback response are also cyborg enhancements. Some theorists cite such modifications as contact lenses, hearing aids, or intraocular lenses as examples of fitting humans with technology to enhance their biological capabilities. As cyborgs currently are on the rise some theorists argue there is a need to develop new definitions of aging and for instance a bio-techno-social definition of aging has been suggested.

The term is also used to address human-technology mixtures in the abstract. This includes not only commonly used pieces of technology such as phones, computers, the Internet, etc. but also artifacts that may not popularly be considered technology; for example, pen and paper, and speech and language. When augmented with these technologies and connected in communication with people in other times and places, a person becomes capable of much more than they were before. This is like a computer, which gains power by using Internet protocols to connect with other computers. Cybernetic technologies include highways, pipes, electrical wiring, buildings, electrical plants, libraries, and other infrastructure that we hardly notice, but which are critical parts of the cybernetics that we work within.

Bruce Sterling in his universe of Shaper/Mechanist suggested an idea of alternative cyborg called Lobster, which is made not by using internal implants, but by using an external shell (*e.g.* a Powered Exoskeleton). Unlike human cyborgs that appear human externally while being synthetic internally (*e.g.* the Bishop type in the Alien franchise), Lobster looks inhuman externally but contains a human internally (*e.g.* Elysium, RoboCop). The computer game *Deus Ex: Invisible War* prominently featured cyborgs called Omar, where "Omar" is a Russian translation of the word "Lobster" (since the Omar are of Russian origin in the game).

Origins

The concept of a man-machine mixture was widespread in science fiction before World War II. As early as 1843, Edgar Allan Poe described a man with extensive prostheses in the short story "The Man That Was Used Up". In 1911, Jean de la Hire introduced the Nyctalope, a science fiction hero who was perhaps the first literary cyborg, in *Le Mystère des XV* (later translated as *The Nyctalope on Mars*). Edmond Hamilton presented space explorers with a mixture of organic and machine parts in his novel *The Comet Doom* in 1928. He later featured the talking, living brain of an old scientist, Simon Wright, floating around in a transparent case, in all the adventures of his famous hero, Captain Future. He uses the term explicitly in the 1962 short story, "After a Judgment Day," to describe the "mechanical analogs" called "Charlies," explaining that "[c] yborgs, they had been called from the first one in the 1960s...cybernetic organisms." In the short story "No Woman Born" in 1944, C. L. Moore wrote of Deirdre, a dancer, whose body was burned completely and whose brain was placed in a faceless but beautiful and supple mechanical body.

The term was coined by Manfred E. Clynes and Nathan S. Kline in 1960 to refer to their conception of an enhanced human being who could survive in extraterrestrial environments:

> "For the exogenously extended organizational complex functioning as an integrated homeostatic system unconsciously, we propose the term 'Cyborg'. – Manfred E. Clynes and Nathan S. Kline."

Their concept was the outcome of thinking about the need for an intimate relationship between human and machine as the new frontier of space exploration was beginning to open up. A designer of physiological instrumentation and electronic data-processing systems, Clynes was the chief research scientist in the Dynamic Simulation Laboratory at Rockland State Hospital in New York.

The term first appears in print five months earlier when *The New York Times* reported on the Psychophysiological Aspects of Space Flight Symposium where Clynes and Kline first presented their paper.

> "A cyborg is essentially a man-machine system in which the control mechanisms of the human portion are modified externally by drugs or regulatory devices so that the being can live in an environment different from the normal one."

A book titled *Cyborg: Digital Destiny and Human Possibility in the Age of the Wearable Computer* was published by Doubleday in 2001. Some of the ideas in the book were incorporated into the 35 mm motion picture film *Cyberman*.

Cyborg Tissues in Engineering

Cyborgs tissues structured with carbon nanotubes and plant or fungal cells have been used in artificial tissue engineering to produce new materials for mechanical and electrical uses. The work was presented by Di Giacomo and Maresca at MRS 2013 Spring conference on Apr, 3rd, talk number SS4.04. The cyborg obtained is inexpensive, light and has unique mechanical properties. It can also be shaped in desired forms. Cells combined with MWCNTs co-precipitated as a specific aggregate of cells and nanotubes that formed a viscous material. Likewise, dried cells still acted as a stable matrix for the MWCNT network. When observed by optical microscopy the material resembled an artificial "tissue" composed of highly packed cells. The effect of cell drying is manifested by their "ghost cell" appearance. A rather specific physical interaction between MWCNTs and cells was observed by electron microscopy suggesting that the cell wall (the most outer part of fungal and plant cells) may play a major active role in establishing a CNTs network and its stabilization. This novel material can be used in a wide range of electronic applications from heating to sensing and has the potential to open important new avenues to be exploited in electromagnetic shielding for radio frequency electronics and aerospace technology. In particular using Candida albicans cells cyborg tissue materials with temperature sensing properties have been reported.

Actual Cyborgization Attempts

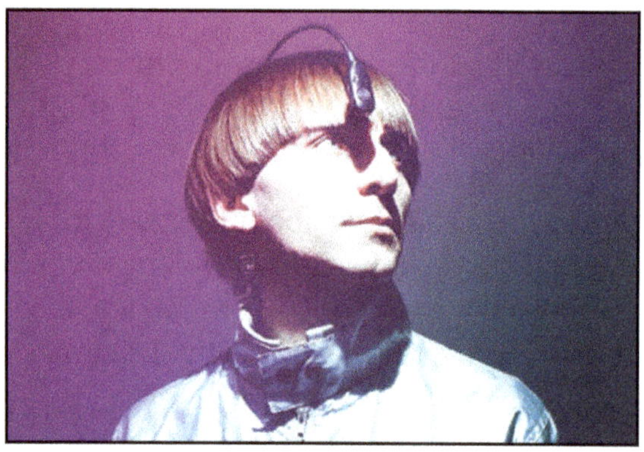

Neil Harbisson, cyborg activist and president of the Cyborg Foundation.

In current prosthetic applications, the C-Leg system developed by Otto Bock HealthCare is used to replace a human leg that has been amputated because of injury or illness. The use of sensors in the artificial C-Leg aids in walking significantly by attempting to replicate the user's natural gait, as it would be prior to amputation. Prostheses like the C-Leg and the more advanced iLimb are considered by some to be the first real steps towards the next generation of real-world cyborg applications. Additionally cochlear implants and magnetic implants which provide people with a sense that they would not otherwise have had can additionally be thought of as creating cyborgs.

In vision science, direct brain implants have been used to treat non-congenital (acquired) blindness. One of the first scientists to come up with a working brain interface to restore sight was private researcher William Dobelle. Dobelle's first prototype was implanted into "Jerry", a man blinded in adulthood, in 1978. A single-array BCI containing 68 electrodes was implanted onto Jerry's visual cortex and succeeded in producing phosphenes, the sensation of seeing light. The system included cameras mounted on glasses to send signals to the implant. Initially, the implant allowed Jerry to see shades of grey in a limited field of vision at a low frame-rate. This also required him to be hooked up to a two-ton mainframe, but shrinking electronics and faster computers made his artificial eye more portable and now enable him to perform simple tasks unassisted.

In 1997, Philip Kennedy, a scientist and physician, created the world's first human cyborg from Johnny Ray, a Vietnam veteran who suffered a stroke. Ray's body, as doctors called it, was "locked in". Ray wanted his old life back so he agreed to Kennedy's experiment. Kennedy embedded an implant he designed (and named "neurotrophic electrode") near the part of Ray's brain so that Ray would be able to have some movement back in his body. The surgery went successfully, but in 2002, Johnny Ray died.

In 2002, Canadian Jens Naumann, also blinded in adulthood, became the first in a series of 16 paying patients to receive Dobelle's second generation implant, marking one of the earliest commercial uses of BCIs. The second generation device used a more sophisticated implant enabling better mapping of phosphenes into coherent vision. Phosphenes are spread out across the visual field in what researchers call the starry-night effect. Immediately after his implant, Jens was able to use his imperfectly restored vision to drive slowly around the parking area of the research institute.

In 2002, under the heading Project Cyborg, a British scientist, Kevin Warwick, had an array of 100 electrodes fired into his nervous system in order to link his nervous system into the internet. With this in place he successfully carried out a series of experiments including extending his nervous system over the internet to control a robotic hand, a loudspeaker and amplifier. This is a form of extended sensory input and the first direct electronic communication between the nervous systems of two humans.

In 2004, under the heading Bridging the Island of the Colourblind Project, a British and completely color-blind artist, Neil Harbisson, started wearing an eyeborg on his head in order to perceive colors through hearing. His prosthetic device was included within his 2004 passport photograph which has been claimed to confirm his cyborg status. In 2012 at TEDGlobal, Harbisson explained that he did not feel like a cyborg when he started to use the eyeborg, he started to feel like a cyborg when he noticed that the software and his brain had united and given him an extra sense.

Animal Cyborgs

The US-based company Backyard Brains released what they refer to as "The world's first commercially available cyborg" called the RoboRoach. The project started as a University of Michigan biomedical engineering student senior design project in 2010 and was launched as an available beta product on 25 February 2011. The RoboRoach was officially released into production via a TED talk at the TED Global conference, and via the crowdsourcing website Kickstarter in 2013, the kit allows students to use microstimulation to momentarily control the movements of a walking cockroach (left and right) using a bluetooth-enabled smartphone as the controller. Other groups have developed cyborg insects, including researchers at North Carolina State University and UC Berkeley, but the RoboRoach was the first kit available to the general public and was funded by the National Institute of Mental Health as a device to serve as a teaching aid to promote an interest in neuroscience. Several animal welfare organizations including the RSPCA and PETA have expressed concerns about the ethics and welfare of animals in this project.

Cyborg Proliferation in Society

In Medicine

In medicine, there are two important and different types of cyborgs: the restorative and the enhanced. Restorative technologies "restore lost function, organs, and limbs". The key aspect of restorative cyborgization is the repair of broken or missing processes to revert to a healthy or average level of function. There is no enhancement to the original faculties and processes that were lost.

On the contrary, the enhanced cyborg "follows a principle, and it is the principle of optimal performance: maximising output (the information or modifications obtained) and minimising input (the energy expended in the process)". Thus, the enhanced cyborg intends to exceed normal processes or even gain new functions that were not originally present.

Although prostheses in general supplement lost or damaged body parts with the integration of a mechanical artifice, bionic implants in medicine allow model organs or body parts to mimic the original function more closely. Michael Chorost wrote a memoir of his experience with cochlear implants, or bionic ear, titled "Rebuilt: How Becoming Part Computer Made Me More Human." Jesse Sullivan became one of the first people to operate a fully robotic limb through a nerve-muscle graft, enabling him a complex range of motions beyond that of previous prosthetics. By 2004, a fully functioning artificial heart was developed. The continued technological development of bionic and nanotechnologies begins to raise the question of enhancement, and of the future possibilities for cyborgs which surpass the original functionality of the biological model. The ethics and desirability of "enhancement prosthetics" have been debated; their proponents include the transhumanist movement, with its belief that new technologies can assist the human race in developing beyond its present, normative limitations such as aging and disease, as well as other, more general incapacities, such as limitations on speed, strength, endurance, and intelligence. Opponents of the concept describe what they believe to be biases which propel the development and acceptance of such technologies; namely, a bias towards functionality and efficiency that may compel assent to a view of human people which de-emphasizes as defining characteristics actual manifestations of humanity and personhood, in favor of definition in terms of upgrades, versions, and utility.

A brain-computer interface, or BCI, provides a direct path of communication from the brain to an external device, effectively creating a cyborg. Research of Invasive BCIs, which utilize electrodes implanted directly into the grey matter of the brain, has focused on restoring damaged eyesight in the blind and providing functionality to paralyzed people, most notably those with severe cases, such as Locked-In syndrome. This technology could enable people who are missing a limb or are in a wheelchair the power to control the devices that aide them through neural signals sent from the brain implants directly to computers or the devices. It is possible that this technology will also eventually be used with healthy people.

Deep brain stimulation is a neurological surgical procedure used for therapeutic purposes. This process has aided in treating patients diagnosed with Parkinson's disease, Alzheimer's disease, Tourette syndrome, epilepsy, chronic headaches, and mental disorders. After the patient is unconscious, through anesthesia, brain pacemakers or electrodes, are implanted into the region of the brain where the cause of the disease is present. The region of the brain is then stimulated by bursts of electric current to disrupt the oncoming surge of seizures. Like all invasive procedures, deep brain stimulation may put the patient at a higher risk. However, there have been more improvements in recent years with deep brain stimulation than any available drug treatment.

Retinal implants are another form of cyborgization in medicine. The theory behind retinal stimulation to restore vision to people suffering from retinitis pigmentosa and vision loss due to aging (conditions in which people have an abnormally low amount of ganglion cells) is that the retinal implant and electrical stimulation would act as a substitute for the missing ganglion cells (cells which connect the eye to the brain.).

While work to perfect this technology is still being done, there have already been major advances in the use of electronic stimulation of the retina to allow the eye to sense patterns of light. A specialized camera is worn by the subject, such as on the frames of their glasses, which converts the image into a pattern of electrical stimulation. A chip located in the user's eye would then electrically stimulate the retina with this pattern by exciting certain nerve endings which transmit the image to the optic centers of the brain and the image would then appear to the user. If technological advances proceed as planned this technology may be used by thousands of blind people and restore vision to most of them.

A similar process has been created to aide people who have lost their vocal cords. This experimental device would do away with previously used robotic sounding voice simulators. The transmission of sound would start with a surgery to redirect the nerve that controls the voice and sound production to a muscle in the neck, where a nearby sensor would be able to pick up its electrical signals. The signals would then move to a processor which would control the timing and pitch of a voice simulator. That simulator would then vibrate producing a multitonal sound which could be shaped into words by the mouth.

An article published in *Nature Materials* in 2012 reported a research on "cyborg tissues" (engineered human tissues with embedded three-dimensional mesh of nanoscale wires), with possible medical implications.

In 2014, researchers from the University of Illinois at Urbana-Champaign and Washington University in St. Louis had developed a device that could keep a heart beating endlessly. By using 3D printing and computer modeling these scientist developed an electronic membrane that could successfully replace pacemakers. The device utilizes a "spider-web like network of sensors and electrodes" to monitor and maintain a normal heart-rate with electrical stimuli. Unlike traditional pacemakers that are similar from patient to patient, the elastic heart glove is made custom by using high-resolution imaging technology. The first prototype was created to fit a rabbit's heart, operating the organ in an oxygen and nutrient-rich solution. The stretchable material and circuits of the apparatus were first constructed by Professor John A. Rogers in which the electrodes are arranged in a s-shape design to allow them to expand and bend without breaking. Although the device is only currently used as a research tool to study changes in heart rate, in the future the membrane may serve as a safeguard from heart attacks.

In the Military

Military organizations' research has recently focused on the utilization of cyborg animals for the purposes of a supposed tactical advantage. DARPA has announced its interest in developing "cyborg insects" to transmit data from sensors implanted into the insect during the pupal stage. The insect's motion would be controlled from a Micro-Electro-Mechanical System (MEMS) and could conceivably survey an environment or detect explosives and gas. Similarly, DARPA is developing a neural implant to remotely control the movement of sharks. The shark's unique senses would then be exploited to provide data feedback in relation to enemy ship movement or underwater explosives.

In 2006, researchers at Cornell University invented a new surgical procedure to implant artificial structures into insects during their metamorphic development. The first insect cyborgs, moths with integrated electronics in their thorax, were demonstrated by the same researchers. The initial success of the techniques has resulted in increased research and the creation of a program called Hybrid-Insect-MEMS, HI-MEMS. Its goal, according to DARPA's Microsystems Technology Office, is to develop "tightly coupled machine-insect interfaces by placing micro-mechanical systems inside the insects during the early stages of metamorphosis".

The use of neural implants has recently been attempted, with success, on cockroaches. Surgically applied electrodes were put on the insect, which were remotely controlled by a human. The results, although sometimes different, basically showed that the cockroach could be controlled by the impulses it received through the electrodes. DARPA is now funding this research because of its obvious beneficial applications to the military and other areas.

In 2009 at the Institute of Electrical and Electronics Engineers (IEEE) Micro-electronic mechanical systems (MEMS) conference in Italy, researchers demonstrated the first "wireless" flying-beetle cyborg. Engineers at the University of California at Berkeley have pioneered the design of a "remote controlled beetle", funded by the DARPA HI-MEMS Program. Filmed evidence of this can be viewed here. This was followed later that year by the demonstration of wireless control of a "lift-assisted" moth-cyborg.

Eventually researchers plan to develop HI-MEMS for dragonflies, bees, rats and pigeons. For the HI-MEMS cybernetic bug to be considered a success, it must fly 100 metres (330 ft) from a starting point, guided via computer into a controlled landing within 5 metres (16 ft) of a specific end point. Once landed, the cybernetic bug must remain in place.

In Sports

In Art

The concept of the cyborg is often associated with science fiction. However, many artists have tried to create public awareness of cybernetic organisms; these can range from paintings to installations. Some artists who create such works are Neil Harbisson, Moon Ribas, Patricia Piccinini, Steve Mann, Orlan, H. R. Giger, Lee Bul, Wafaa Bilal, Tim Hawkinson and Stelarc.

Stelarc is a performance artist who has visually probed and acoustically amplified his body. He uses medical instruments, prosthetics, robotics, virtual reality systems, the Internet and biotechnology to explore alternate, intimate and involuntary interfaces with the body. He has made three films of the inside of his body and has performed with a third hand and a virtual arm. Between 1976–1988 he completed 25 body suspension performances with hooks into the skin. For 'Third Ear' he surgically constructed an extra ear within his arm that was internet enabled, making it a publicly accessible acoustical organ for people in other places. He is presently performing as his avatar from his second life site.

Tim Hawkinson promotes the idea that bodies and machines are coming together as one, where human features are combined with technology to create the Cyborg. Hawkinson's piece *Emoter* presented how society is now dependent on technology.

Wafaa Bilal is an Iraqi-American performance artist who had a small 10 megapixel digital camera surgically implanted into the back of his head, part of a project entitled 3rd I. For one year, beginning 15 December 2010, an image is captured once per minute 24 hours a day and streamed live to www.3rdi.me and the Mathaf: Arab Museum of Modern Art. The site also displays Bilal's location via GPS. Bilal says that the reason why he put the camera in the back of the head was to make an "allegorical statement about the things we don't see and leave behind." As a professor at NYU, this project has raised privacy issues, and so Bilal has been asked to ensure that his camera does not take photographs in NYU buildings.

Machines are becoming more ubiquitous in the artistic process itself, with computerized drawing pads replacing pen and paper, and drum machines becoming nearly as popular as human drummers. This is perhaps most notable in generative art and music. Composers such as Brian Eno have developed and utilized software which can build entire musical scores from a few basic mathematical parameters.

Scott Draves is a generative artist whose work is explicitly described as a "cyborg mind". His Electric Sheep project generates abstract art by combining the work of many computers and people over the internet.

Artists as Cyborgs

Artists have explored the term cyborg from a perspective involving imagination. Some work to make an abstract idea of technological and human-bodily union apparent to reality in an art form utilizing varying mediums, from sculptures and drawings to digital renderings. Artists that seek to make cyborg-based fantasies a reality often call themselves cyborg artists, or may consider their artwork "cyborg". How an artist or their work may be considered cyborg will vary depending upon the interpreter's flexibility with the term. Scholars that rely upon a strict, technical description of cyborg, often going by Norbert Wiener's cybernetic theory and Manfred E. Clynes and Nathan S. Kline's first use of the term, would likely argue that most cyborg artists do not qualify to be considered cyborgs. Scholars considering a more flexible description of cyborgs may argue it incorporates more than cybernetics. Others may speak of defining subcategories, or specialized cyborg types, that qualify different levels of cyborg at which technology influences an individual. This may range from technological instruments being external, temporary, and removable to being fully integrated and permanent. Nonetheless, cyborg artists are artists. Being so, it can be expected for them to incorporate the cyborg idea rather than a strict, technical representation of the term, seeing how their work will sometimes revolve around other purposes outside of cyborgism.

In Body Modification

As medical technology becomes more advanced, some techniques and innovations are adopted by the body modification community. While not yet cyborgs in the strict definition of Manfred Clynes and Nathan Kline, technological developments like implantable silicon silk electronics, augmented reality and QR codes are bridging the disconnect between technology and the body. Hypothetical technologies such as digital tattoo interfaces would blend body modification aesthetics with interactivity and functionality, bringing a transhumanist way of life into present day reality.

In addition, it is quite plausible for anxiety expression to manifest. Individuals may experience pre-implantation feelings of fear and nervousness. To this end, individuals may also embody feelings of uneasiness, particularly in a socialized setting, due to their post-operative, technologically augmented bodies, and mutual unfamiliarity with the mechanical insertion. Anxieties may be linked to notions of otherness or a cyborged identity.

In Popular Culture

Cyborgs have become a well-known part of science fiction literature and other media. Although many of these characters may be technically androids, they are often referred to as cyborgs. Well-known examples from film and television include RoboCop, The Terminator, Evangelion, United States Air Force Colonel Steve Austin in both *Cyborg* and, as acted out by Lee Majors, *The Six Million Dollar Man,* Replicants from *Blade Runner*, Daleks and Cybermen from *Doctor Who,* the Borg from *Star Trek,* Darth Vader and General Grievous from *Star Wars*, Inspector Gadget, and Cylons from the 2004 *Battlestar Galactica* series. From manga and anime are characters such as 8 Man (the inspiration for *RoboCop*), Kamen Rider, *Ghost in the Shell's* Motoko Kusanagi, as well as characters from western comic books like Tony Stark (after his Extremis and Bleeding Edge armor) and Victor "Cyborg" Stone. The *Deus Ex* videogame series deals extensively with the near-future rise of cyborgs and their corporate ownership, as does the *Syndicate* series. William Gibson's Neuromancer features one of the first female cyborgs, a "Razorgirl" named Molly Millions, who has extensive cybernetic modifications and is one of the most prolific cyberpunk characters in the science fiction canon.

In Space

Sending humans to space is a dangerous task in which the implementation of various cyborg technologies could be used in the future for risk mitigation. Stephen Hawking, a renowned physicist, stated "Life on Earth is at the ever-increasing risk of being wiped out by a disaster such as sudden global warming, nuclear war... I think the human race has no future if it doesn't go into space." The difficulties associated with space travel could mean it might be centuries before humans ever become a multi-planet species. There are many effect of spaceflight on the human body. One major issue of space exploration is the biological need for oxygen. If this necessity was taken out of the equation, space exploration would be revolutionized. A theory proposed by Manfred E. Clynes and Nathan S. Kline is aimed at tackling this problem. The two scientists theorized that the use of an inverse fuel cell that is "capable of reducing CO_2 to its components with removal of the carbon and re-circulation of the oxygen..." could make breathing unnecessary. Another prominent issue is radiation exposure. Yearly, the average human on earth is exposed to approximately 0.30 rem of radiation, while an astronaut aboard the International Space Station for 90 days is exposed to 9 rem. To tackle the issue, Clynes and Kline theorized a cyborg containing a sensor that would detect radiation levels and a Rose osmotic pump "which would automatically inject protective pharmaceuticals in appropriate doses." Experiments injecting these protective pharmaceuticals into monkeys have shown positive results in increasing radiation resistance.

Although the effects of spaceflight on our body is an important issue, the advancement of propulsion technology is just as important. With our current technology, it would take us about 260 days to get to Mars. A study backed by NASA proposes an interesting way to tackle this issue through deep sleep, or torpor. With this technique, it would "reduce astronauts' metabolic functions with existing medical procedures". So far experiments have only resulted in patients being in torpor state for one week. Advancements to allow for longer states of deep sleep would lower the cost of the trip to mars as a result of reduced astronaut resource consumption.

Cyborgization in Critical Deaf Studies

Joseph Michael Valente, describes "cyborgization" as an attempt to codify "normalization" through cochlear implantation in young deaf children. Drawing from Paddy Ladd's work on Deaf epistemology and Donna Haraway's Cyborg ontology, Valente "use[s] the concept of the cyborg as a way of agitating constructions of cyborg perfection (for the deaf child that would be to become fully hearing)". He claims that cochlear implant manufacturers advertise and sell cochlear implants as a mechanical device as well as an uncomplicated medical "miracle cure". Valente criticizes cochlear implant researchers whose studies largely to date do not include cochlear implant recipients, despite cochlear implants having been approved by the United States Food and Drug Administration (FDA) since 1984. Pamela J. Kincheloe discusses the representation of the cochlear implant in media and popular culture as a case study for present and future responses to human alteration and enhancement.

Cyborg Foundation

In 2010, the Cyborg Foundation became the world's first international organization dedicated to help humans become cyborgs. The foundation was created by cyborg Neil Harbisson and Moon Ribas as a response to the growing amount of letters and emails received from people around the world interested in becoming a cyborg. The foundation's main aims are to extend human senses and abilities by creating and applying cybernetic extensions to the body, to promote the use of cybernetics in cultural events and to defend cyborg rights. In 2010, the foundation, based in Mataró (Barcelona), was the overall winner of the Cre@tic Awards, organized by Tecnocampus Mataró.

In 2012, Spanish film director Rafel Duran Torrent, created a short film about the Cyborg Foundation. In 2013, the film won the Grand Jury Prize at the Sundance Film Festival's Focus Forward Filmmakers Competition and was awarded with $100,000 USD.

Laws of Robotics

Laws of Robotics are a set of laws, rules, or principles, which are intended as a fundamental framework to underpin the behavior of robots designed to have a degree of autonomy. Robots of this degree of complexity do not yet exist, but they have been widely anticipated in science fiction, films and are a topic of active research and development in the fields of robotics and artificial intelligence.

The best known set of laws are those written by Isaac Asimov in the 1940s, or based upon them, but other sets of laws have been proposed by researchers in the decades since then.

Isaac Asimov's "Three Laws of Robotics"

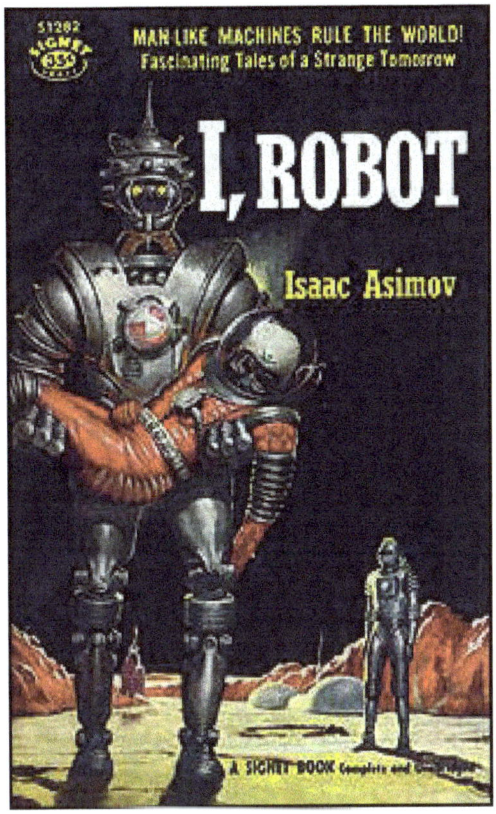

This cover of *I, Robot* illustrates the story "Runaround", the first to list all Three Laws of Robotics.

The best known set of laws are Isaac Asimov's "Three Laws of Robotics". These were introduced in his 1942 short story "Runaround", although they were foreshadowed in a few earlier stories. The Three Laws are:

1. A robot may not injure a human being or, through inaction, allow a human being to come to harm.
2. A robot must obey the orders given it by human beings, except where such orders would conflict with the First Law.
3. A robot must protect its own existence as long as such protection does not conflict with the First or Second Laws.

Near the end of his book *Foundation and Earth*, a zeroth law was introduced:

> 0. A robot may not injure humanity, or, by inaction, allow humanity to come to harm.

Adaptations and extensions exist based upon this framework. As of 2011 they remain a "fictional device".

EPRSC / AHRC Principles of Robotics

In 2011, the Engineering and Physical Sciences Research Council (EPSRC) and the Arts and Humanities Research Council (AHRC) of Great Britain jointly published a set of five ethical "principles for designers, builders and users of robots" in the real world, along with seven "high-level messages" intended to be conveyed, based on a September 2010 research workshop:

1. Robots should not be designed solely or primarily to kill or harm humans.
2. Humans, not robots, are responsible agents. Robots are tools designed to achieve human goals.
3. Robots should be designed in ways that assure their safety and security.
4. Robots are artifacts; they should not be designed to exploit vulnerable users by evoking an emotional response or dependency. It should always be possible to tell a robot from a human.
5. It should always be possible to find out who is legally responsible for a robot.

The messages intended to be conveyed were:

1. We believe robots have the potential to provide immense positive impact to society. We want to encourage responsible robot research.
2. Bad practice hurts us all.
3. Addressing obvious public concerns will help us all make progress.
4. It is important to demonstrate that we, as roboticists, are committed to the best possible standards of practice.
5. To understand the context and consequences of our research, we should work with experts from other disciplines, including: social sciences, law, philosophy and the arts.
6. We should consider the ethics of transparency: are there limits to what should be openly available?
7. When we see erroneous accounts in the press, we commit to take the time to contact the reporting journalists.

Judicial Development

Another comprehensive terminological codification for the legal assessment of the technological developments in the robotics industry has already begun mainly in Asian countries. This progress represents a contemporary reinterpretation of the law (and ethics) in the field of robotics, an interpretation that assumes a rethinking of traditional legal constellations. These include primarily legal liability issues in civil and criminal law.

Satya Nadella's Laws

In June 2016, Satya Nadella, a CEO of Microsoft Corporation at the time, had an interview with the *Slate* magazine and roughly sketched five rules for artificial intelligences to be observed by their designers:

1. "A.I. must be designed to assist humanity" meaning human autonomy needs to be respected.
2. "A.I. must be transparent" meaning that humans should know and be able to understand how they work.
3. "A.I. must maximize efficiencies without destroying the dignity of people".
4. "A.I. must be designed for intelligent privacy" meaning that it earns trust through guarding their information.
5. "A.I. must have algorithmic accountability so that humans can undo unintended harm".
6. "A.I. must guard against bias" so that they must not discriminate people.

Tilden's "Laws of Robotics"

Mark W. Tilden proposed three guiding principles/rules for robots, which do not pertain to humans or humanity, but to robots themselves:

1. A robot must protect its existence at all costs.
2. A robot must obtain and maintain access to its own power source.
3. A robot must continually search for better power sources.

Three Laws of Robotics

The Three Laws of Robotics (often shortened to The Three Laws or known as Asimov's

Laws) are a set of rules devised by the science fiction author Isaac Asimov. The rules were introduced in his 1942 short story "Runaround", although they had been foreshadowed in a few earlier stories. The Three Laws, quoted as being from the "Handbook of Robotics, 56th Edition, 2058 A.D.", are:

1. A robot may not injure a human being or, through inaction, allow a human being to come to harm.

2. A robot must obey the orders given [*sic*] it by human beings except where such orders would conflict with the First Law.

3. A robot must protect its own existence as long as such protection does not conflict with the First or Second Laws.

These form an organizing principle and unifying theme for Asimov's robotic-based fiction, appearing in his *Robot* series, the stories linked to it, and his *Lucky Starr* series of young-adult fiction. The Laws are incorporated into almost all of the positronic robots appearing in his fiction, and cannot be bypassed, being intended as a safety feature. Many of Asimov's robot-focused stories involve robots behaving in unusual and counter-intuitive ways as an unintended consequence of how the robot applies the Three Laws to the situation in which it finds itself. Other authors working in Asimov's fictional universe have adopted them and references, often parodic, appear throughout science fiction as well as in other genres.

The original laws have been altered and elaborated on by Asimov and other authors. Asimov himself made slight modifications to the first three in various books and short stories to further develop how robots would interact with humans and each other. In later fiction where robots had taken responsibility for government of whole planets and human civilizations, Asimov also added a fourth, or zeroth law, to precede the others:

0. A robot may not harm humanity, or, by inaction, allow humanity to come to harm.

The Three Laws, and the zeroth, have pervaded science fiction and are referred to in many books, films, and other media.

History

In *The Rest of the Robots*, published in 1964, Asimov noted that when he began writing in 1940 he felt that "one of the stock plots of science fiction was ... robots were created and destroyed their creator. Knowledge has its dangers, yes, but is the response to be a retreat from knowledge? Or is knowledge to be used as itself a barrier to the dangers it brings?" He decided that in his stories robots would not "turn stupidly on his creator for no purpose but to demonstrate, for one more weary time, the crime and punishment of Faust."

On May 3, 1939, Asimov attended a meeting of the Queens (New York) Science Fiction Society where he met Ernest and Otto Binder who had recently published a short story "I, Robot" featuring a sympathetic robot named Adam Link who was misunderstood and motivated by love and honor. (This was the first of a series of ten stories; the next year "Adam Link's Vengeance" (1940) featured Adam thinking "A robot must never kill a human, of his own free will.") Asimov admired the story. Three days later Asimov began writing "my own story of a sympathetic and noble robot", his 14th story. Thirteen days later he took "Robbie" to John W. Campbell the editor of *Astounding Science-Fiction*. Campbell rejected it claiming that it bore too strong a resemblance to Lester del Rey's "Helen O'Loy", published in December 1938; the story of a robot that is so much like a person that she falls in love with her creator and becomes his ideal wife. Frederik Pohl published "Robbie" in *Astonishing Stories* magazine the following year.

Asimov attributes the Three Laws to John W. Campbell, from a conversation that took place on 23 December 1940. Campbell claimed that Asimov had the Three Laws already in his mind and that they simply needed to be stated explicitly. Several years later Asimov's friend Randall Garrett attributed the Laws to a symbiotic partnership between the two men – a suggestion that Asimov adopted enthusiastically. According to his autobiographical writings Asimov included the First Law's "inaction" clause because of Arthur Hugh Clough's poem "The Latest Decalogue", which includes the satirical lines "Thou shalt not kill, but needst not strive / officiously to keep alive".

Although Asimov pins the creation of the Three Laws on one particular date, their appearance in his literature happened over a period. He wrote two robot stories with no explicit mention of the Laws, "Robbie" and "Reason". He assumed, however, that robots would have certain inherent safeguards. "Liar!", his third robot story, makes the first mention of the First Law but not the other two. All three laws finally appeared together in "Runaround". When these stories and several others were compiled in the anthology *I, Robot*, "Reason" and "Robbie" were updated to acknowledge all the Three Laws, though the material Asimov added to "Reason" is not entirely consistent with the Three Laws as he described them elsewhere. In particular the idea of a robot protecting human lives when it does not believe those humans truly exist is at odds with Elijah Baley's reasoning, as described below.

During the 1950s Asimov wrote a series of science fiction novels expressly intended for young-adult audiences. Originally his publisher expected that the novels could be adapted into a long-running television series, something like *The Lone Ranger* had been for radio. Fearing that his stories would be adapted into the "uniformly awful" programming he saw flooding the television channels Asimov decided to publish the *Lucky Starr* books under the pseudonym "Paul French". When plans for the television series fell through, Asimov decided to abandon the pretence; he brought the Three Laws into *Lucky Starr and the Moons of Jupiter*, noting that this "was a dead giveaway to Paul French's identity for even the most casual reader".

In his short story "Evidence" Asimov lets his recurring character Dr. Susan Calvin expound a moral basis behind the Three Laws. Calvin points out that human beings are typically expected to refrain from harming other human beings (except in times of extreme duress like war, or to save a greater number) and this is equivalent to a robot's First Law. Likewise, according to Calvin, society expects individuals to obey instructions from recognized authorities such as doctors, teachers and so forth which equals the Second Law of Robotics. Finally humans are typically expected to avoid harming themselves which is the Third Law for a robot.

The plot of "Evidence" revolves around the question of telling a human being apart from a robot constructed to appear human – Calvin reasons that if such an individual obeys the Three Laws he may be a robot or simply "a very good man". Another character then asks Calvin if robots are very different from human beings after all. She replies, "Worlds different. Robots are essentially decent."

Asimov later wrote that he should not be praised for creating the Laws, because they are "obvious from the start, and everyone is aware of them subliminally. The Laws just never happened to be put into brief sentences until I managed to do the job. The Laws apply, as a matter of course, to every tool that human beings use", and "analogues of the Laws are implicit in the design of almost all tools, robotic or not":

1. Law 1: A tool must not be unsafe to use. Hammers have handles and screwdrivers have hilts to help increase grip. It is of course possible for a person to injure himself with one of these tools, but that injury would only be due to his incompetence, not the design of the tool.

2. Law 2: A tool must perform its function efficiently unless this would harm the user. This is the entire reason ground-fault circuit interrupters exist. Any running tool will have its power cut if a circuit senses that some current is not returning to the neutral wire, and hence might be flowing through the user. The safety of the user is paramount.

3. Law 3: A tool must remain intact during its use unless its destruction is required for its use or for safety. For example, Dremel disks are designed to be as tough as possible without breaking unless the job requires it to be spent. Furthermore, they are designed to break at a point before the shrapnel velocity could seriously injure someone (other than the eyes, though safety glasses should be worn at all times anyway).

Asimov believed that, ideally, humans would also follow the Laws:

I have my answer ready whenever someone asks me if I think that my Three Laws of Robotics will actually be used to govern the behavior of robots, once they become versatile and flexible enough to be able to choose among different courses of behavior.

My answer is, "Yes, the Three Laws are the only way in which rational human beings can deal with robots—or with anything else."

—But when I say that, I always remember (sadly) that human beings are not always rational.

Alterations

By Asimov

Asimov's stories test his Three Laws in a wide variety of circumstances leading to proposals and rejection of modifications. Science fiction scholar James Gunn writes in 1982, "The Asimov robot stories as a whole may respond best to an analysis on this basis: the ambiguity in the Three Laws and the ways in which Asimov played twenty-nine variations upon a theme". While the original set of Laws provided inspirations for many stories, Asimov introduced modified versions from time to time.

First Law Modified

In "Little Lost Robot" several NS-2, or "Nestor", robots are created with only part of the First Law. It reads:

1. A robot may not harm a human being.

This modification is motivated by a practical difficulty as robots have to work alongside human beings who are exposed to low doses of radiation. Because their positronic brains are highly sensitive to gamma rays the robots are rendered inoperable by doses reasonably safe for humans. The robots are being destroyed attempting to rescue the humans who are in no actual danger but "might forget to leave" the irradiated area within the exposure time limit. Removing the First Law's "inaction" clause solves this problem but creates the possibility of an even greater one: a robot could initiate an action that would harm a human (dropping a heavy weight and failing to catch it is the example given in the text), knowing that it was capable of preventing the harm and then decide not to do so.

Gaia is a planet with collective intelligence in the *Foundation* which adopts a law similar to the First Law, and the Zeroth Law, as its philosophy:

Gaia may not harm life or allow life to come to harm.

Zeroth Law Added

Asimov once added a "Zeroth Law"—so named to continue the pattern where lower-numbered laws supersede the higher-numbered laws—stating that a robot must not harm humanity. The robotic character R. Daneel Olivaw was the first to give the Zeroth Law a name in the novel *Robots and Empire*; however, the character Susan Calvin articulates the concept in the short story "The Evitable Conflict".

In the final scenes of the novel *Robots and Empire*, R. Giskard Reventlov is the first robot to act according to the Zeroth Law. Giskard is telepathic, like the robot Herbie in the short story "Liar!", and tries to apply the Zeroth Law through his understanding of a more subtle concept of "harm" than most robots can grasp. However, unlike Herbie, Giskard grasps the philosophical concept of the Zeroth Law allowing him to harm individual human beings if he can do so in service to the abstract concept of humanity. The Zeroth Law is never programmed into Giskard's brain but instead is a rule he attempts to comprehend through pure metacognition. Though he fails – it ultimately destroys his positronic brain as he is not certain whether his choice will turn out to be for the ultimate good of humanity or not – he gives his successor R. Daneel Olivaw his telepathic abilities. Over the course of many thousands of years Daneel adapts himself to be able to fully obey the Zeroth Law. As Daneel formulates it, in the novels *Foundation and Earth* and *Prelude to Foundation*, the Zeroth Law reads:

A robot may not harm humanity, or, by inaction, allow humanity to come to harm.

A condition stating that the Zeroth Law must not be broken was added to the original Three Laws, although Asimov recognized the difficulty such a law would pose in practice.

Trevize frowned. "How do you decide what is injurious, or not injurious, to humanity as a whole?"

"Precisely, sir," said Daneel. "In theory, the Zeroth Law was the answer to our problems. In practice, we could never decide. A human being is a concrete object. Injury to a person can be estimated and judged. Humanity is an abstraction."

—*Foundation and Earth*

A translator incorporated the concept of the Zeroth Law into one of Asimov's novels before Asimov himself made the law explicit. Near the climax of *The Caves of Steel*, Elijah Baley makes a bitter comment to himself thinking that the First Law forbids a robot from harming a human being. He determines that it must be so unless the robot is clever enough to comprehend that its actions are for humankind's long-term good. In Jacques Brécard's 1956 French translation entitled *Les Cavernes d'acier* Baley's thoughts emerge in a slightly different way:

"A robot may not harm a human being, unless he finds a way to prove that ultimately the harm done would benefit humanity in general!"

Removal of the Three Laws

Asimov portrayed robots that disregard the Three Laws entirely thrice during his writing career. The first case was a short-short story entitled "First Law" and is often considered an insignificant "tall tale" or even apocryphal. On the other hand, the short story "Cal" (from the collection *Gold*), told by a first-person robot narrator, features a robot who disregards the Three Laws because he has found something far more important—he wants to be a writer. Humorous, partly autobiographical and unusually experimental in style, "Cal" has been regarded as one of *Gold's* strongest stories. The third is a short story entitled "Sally" in which cars fitted with positronic brains are apparently able to harm and kill humans in disregard of the First Law. However, aside from the positronic brain concept, this story does not refer to other robot stories and may not be set in the same continuity.

The title story of the *Robot Dreams* collection portrays LVX-1, or "Elvex", a robot who enters a state of unconsciousness and dreams thanks to the unusual fractal construction of his positronic brain. In his dream the first two Laws are absent and the Third Law reads "A robot must protect its own existence".

Asimov took varying positions on whether the Laws were optional: although in his first writings they were simply carefully engineered safeguards, in later stories Asimov stated that they were an inalienable part of the mathematical foundation underlying the positronic brain. Without the basic theory of the Three Laws the fictional scientists of Asimov's universe would be unable to design a workable brain unit. This is historically consistent: the occasions where roboticists modify the Laws generally occur early within the stories' chronology and at a time when there is less existing work to be re-done. In "Little Lost Robot" Susan Calvin considers modifying the Laws to be a terrible idea, although possible, while centuries later Dr. Gerrigel in *The Caves of Steel* believes it to be impossible.

The character Dr. Gerrigel uses the term "Asenion" to describe robots programmed with the Three Laws. The robots in Asimov's stories, being Asenion robots, are incapable of knowingly violating the Three Laws but, in principle, a robot in science fiction or in the real world could be non-Asenion. "Asenion" is a misspelling of the name Asimov which was made by an editor of the magazine *Planet Stories*. Asimov used this obscure variation to insert himself into *The Caves of Steel* just like he referred to himself as "Azimuth or, possibly, Asymptote" in *Thiotimoline to the Stars*, in much the same way that Vladimir Nabokov appeared in *Lolita* anagrammatically disguised as "Vivian Darkbloom".

Characters within the stories often point out that the Three Laws, as they exist in a robot's mind, are not the written versions usually quoted by humans but abstract mathematical concepts upon which a robot's entire developing consciousness is based. This concept is largely fuzzy and unclear in earlier stories depicting very rudimentary robots who are only programmed to comprehend basic physical tasks, where the Three Laws act as an overarching safeguard, but by the era of *The Caves of Steel* featuring robots with human or beyond-human intelligence the Three Laws have become the underlying basic ethical worldview that determines the actions of all robots.

By other Authors

Roger MacBride Allen's Trilogy

In the 1990s, Roger MacBride Allen wrote a trilogy which was set within Asimov's fictional universe. Each title has the prefix "Isaac Asimov's" as Asimov had approved Allen's outline before his death. These three books, *Caliban, Inferno* and *Utopia*, introduce a new set of the Three Laws. The so-called New Laws are similar to Asimov's originals with the following differences: the First Law is modified to remove the "inaction" clause, the same modification made in "Little Lost Robot"; the Second Law is modified to require cooperation instead of obedience; the Third Law is modified so it is no longer superseded by the Second (i.e., a "New Law" robot cannot be ordered to destroy itself); finally, Allen adds a Fourth Law which instructs the robot to do "whatever it likes" so long as this does not conflict with the first three laws. The philosophy behind these changes is that "New Law" robots should be partners rather than slaves to humanity, according to Fredda Leving, who designed these New Law Robots. According to the first book's introduction, Allen devised the New Laws in discussion with Asimov himself. However, the *Encyclopedia of Science Fiction* says that "With permission from Asimov, Allen rethought the Three Laws and developed a new set".

Jack Williamson's with Folded Han*ds*

Jack Williamson's novelette *With Folded Hands* (1947), later rewritten as the novel *The Humanoids,* deals with robot servants whose prime directive is "To Serve and Obey, And Guard Men From Harm." While Asimov's robotic laws are meant to protect humans from harm, the robots in Williamson's story have taken these instructions to the extreme; they protect humans from everything, including unhappiness, stress, unhealthy lifestyle and all actions that could be potentially dangerous. All that is left for humans to do is to sit with folded hands.

Foundation Sequel Trilogy

In the officially licensed *Foundation* sequels *Foundation's Fear, Foundation and Chaos* and *Foundation's Triumph* (by Gregory Benford, Greg Bear and David Brin respectively) the future Galactic Empire is seen to be controlled by a conspiracy of humaniform robots who follow the Zeroth Law and led by R. Daneel Olivaw.

The Laws of Robotics are portrayed as something akin to a human religion, and referred to in the language of the Protestant Reformation, with the set of laws containing the Zeroth Law known as the "Giskardian Reformation" to the original "Calvinian Orthodoxy" of the Three Laws. Zeroth-Law robots under the control of R. Daneel Olivaw are seen continually struggling with "First Law" robots who deny the existence of the Zeroth Law, promoting agendas different from Daneel's. Some of these agendas are based on the first clause of the First Law ("A robot may not injure a human being...") advocating strict non-interference in human politics to avoid unwittingly causing harm. Others are based on the second clause ("...or, through inaction, allow a human being to come to harm") claiming that robots should openly become a dictatorial government to protect humans from all potential conflict or disaster.

Daneel also comes into conflict with a robot known as R. Lodovic Trema whose positronic brain was infected by a rogue AI — specifically, a simulation of the long-dead Voltaire — which consequently frees Trema from the Three Laws. Trema comes to believe that humanity should be free to choose its own future. Furthermore, a small group of robots claims that the Zeroth Law of Robotics itself implies a higher Minus One Law of Robotics:

A robot may not harm sentience or, through inaction, allow sentience to come to harm.

They therefore claim that it is morally indefensible for Daneel to ruthlessly sacrifice robots and extraterrestrial sentient life for the benefit of humanity. None of these reinterpretations successfully displace Daneel's Zeroth Law — though *Foundation's Triumph* hints that these robotic factions remain active as fringe groups up to the time of the novel *Foundation*.

These novels take place in a future dictated by Asimov to be free of obvious robot presence and surmise that R. Daneel's secret influence on history through the millennia has prevented both the rediscovery of positronic brain technology and the opportunity to work on sophisticated intelligent machines. This lack of rediscovery and lack of opportunity makes certain that the superior physical and intellectual power wielded by intelligent machines remains squarely in the possession of robots obedient to some form of the Three Laws. That R. Daneel is not entirely successful at this becomes clear in a brief period when scientists on Trantor develop "*tiktoks*" — simplistic programmable machines akin to real–life modern robots and therefore lacking the Three Laws. The robot conspirators see the Trantorian tiktoks as a massive threat to social stability, and their plan to eliminate the tiktok threat forms much of the plot of *Foundation's Fear*.

In *Foundation's Triumph* different robot factions interpret the Laws in a wide variety of ways, seemingly ringing every possible permutation upon the Three Laws' ambiguities.

Robot Mystery Series

Set between *The Robots of Dawn* and *Robots and Empire*, Mark W. Tiedemann's *Robot Mystery* trilogy updates the *Robot–Foundation* saga with robotic minds housed in computer mainframes rather than humanoid bodies. The 2002 Aurora novel has robotic characters debating the moral implications of harming cyborg lifeforms who are part artificial and part biological.

One should not neglect Asimov's own creations in these areas such as the Solarian "viewing" technology and the machines of *The Evitable Conflict* originals that Tiedemann acknowledges. *Aurora*, for example, terms the Machines "the first RIs, really". In addition the *Robot Mystery* series addresses the problem of nanotechnology: building a positronic brain capable of reproducing human cognitive processes requires a high degree of miniaturization, yet Asimov's stories largely overlook the effects this miniaturization would have in other fields of technology. For example, the police department card-readers in *The Caves of Steel* have a capacity of only a few kilobytes per square centimeter of storage medium. *Aurora*, in particular, presents a sequence of historical developments which explains the lack of nanotechnology — a partial retcon, in a sense, of Asimov's timeline.

Additional Laws

There are three Fourth Laws written by authors other than Asimov. The 1974 Lyuben Dilov novel, *Icarus's Way* (a.k.a., *The Trip of Icarus*) introduced a Fourth Law of robotics:

A robot must establish its identity as a robot in all cases.

Dilov gives reasons for the fourth safeguard in this way: "The last Law has put an end to the expensive aberrations of designers to give psychorobots as humanlike a form as possible. And to the resulting misunderstandings..."

A fifth law was introduced by Nikola Kesarovski in his short story "The Fifth Law of Robotics". This fifth law says:

A robot must know it is a robot.

The plot revolves around a murder where the forensic investigation discovers that the victim was killed by a hug from a humaniform robot. The robot violated both the First Law and Dilov's Fourth Law (assumed in Kesarovksi's universe to be the valid one) because it did not establish for itself that it was a robot. The story was reviewed by Valentin D. Ivanov in SFF review webzine The Portal.

For the 1986 tribute anthology, *Foundation's Friends,* Harry Harrison wrote a story entitled, "The Fourth Law of Robotics". This Fourth Law states:

A robot must reproduce. As long as such reproduction does not interfere with the First or Second or Third Law.

In the book a robot rights activist, in an attempt to liberate robots, builds several equipped with this Fourth Law. The robots accomplish the task laid out in this version of the Fourth Law by building new robots who view their creator robots as parental figures.

In reaction to the 2004 Will Smith film adaptation of *I, Robot*, humorist and graphic designer Mark Sottilaro farcically declared the Fourth Law of Robotics to be "When turning evil, display a red indicator light." The red light indicated the wireless uplink to the manufacturer is active, first seen during a software update and later on "Evil" robots taken over by the manufacturer's positronic superbrain.

In 2013 Hutan Ashrafian, proposed an additional law that for the first time considered the role of artificial intelligence-on-artificial intelligence or the relationship between robots themselves – the so-called AIonAI law. This sixth law states:

All robots endowed with comparable human reason and conscience should act towards one another in a spirit of brotherhood.

In Karl Schroeder's *Lockstep* (2014) a character reflects that robots "probably had multiple layers of programming to keep [them] from harming anybody. Not three laws, but twenty or thirty."

Ambiguities and Loopholes

Unknowing Breach of the Laws

In *The Naked Sun*, Elijah Baley points out that the Laws had been deliberately misrepresented because robots could *unknowingly* break any of them. He restated the first law as "A robot may do nothing that, *to its knowledge,* will harm a human being; nor, through inaction, *knowingly* allow a human being to come to harm." This change in wording makes it clear that robots can become the tools of murder, provided they not be aware of the nature of their tasks; for instance being ordered to add something to a person's food, not knowing that it is poison. Furthermore, he points out that a clever criminal could divide a task among multiple robots so that no individual robot could recognize that its actions would lead to harming a human being. *The Naked Sun* complicates the issue by portraying a decentralized, planetwide communication network among Solaria's millions of robots meaning that the criminal mastermind could be located anywhere on the planet.

Baley furthermore proposes that the Solarians may one day use robots for military purposes. If a spacecraft was built with a positronic brain and carried neither humans nor the life-support systems to sustain them, then the ship's robotic intelligence could naturally assume that all other spacecraft were robotic beings. Such a ship could operate more responsively and flexibly than one crewed by humans, could be armed more heavily and its robotic brain equipped to slaughter humans of whose existence it is totally ignorant. This possibility is referenced in *Foundation and Earth* where it is discovered that the Solarians possess a strong police force of unspecified size that has been programmed to identify only the Solarian race as human.

Ambiguities Resulting from Lack of Definition

The Laws of Robotics presume that the terms "human being" and "robot" are understood and well defined. In some stories this presumption is overturned.

Definition of "Human Being"

The Solarians create robots with the Three Laws but with a warped meaning of "human". Solarian robots are told that only people speaking with a Solarian accent are human. This enables their robots to have no ethical dilemma in harming non-Solarian human beings (and are specifically programmed to do so). By the time period of *Foundation and Earth* it is revealed that the Solarians have genetically modified themselves into a distinct species from humanity – becoming hermaphroditic and telekinetic and containing biological organs capable of individually powering and controlling whole complexes of robots. The robots of Solaria thus respected the Three Laws only with regard to the "humans" of Solaria. It is unclear whether all the robots had such definitions, since only the overseer and guardian robots were shown explicitly to have them. In "Robots and Empire", the lower class robots were instructed by their overseer about whether certain creatures are human or not.

Asimov addresses the problem of humanoid robots ("androids" in later parlance) several times. The novel *Robots and Empire* and the short stories "Evidence" and "The Tercentenary Incident" describe robots crafted to fool people into believing that the robots are human. On the other hand, "The Bicentennial Man" and "—That Thou art Mindful of Him" explore how the robots may change their interpretation of the Laws as they grow more sophisticated. Gwendoline Butler writes in *A Coffin for the Canary* "Perhaps we are robots. Robots acting out the last Law of Robotics... To tend towards the human." In *The Robots of Dawn*, Elijah Baley points out that the use of humaniform robots as the first wave of settlers on new Spacer worlds may lead to the robots seeing themselves as the true humans, and deciding to keep the worlds for themselves rather than allow the Spacers to settle there.

"—That Thou art Mindful of Him", which Asimov intended to be the "ultimate" probe into the Laws' subtleties, finally uses the Three Laws to conjure up the very "Frankenstein" scenario they were invented to prevent. It takes as its concept the growing development of robots that mimic non-human living things and given programs that mimic simple animal behaviours which do not require the Three Laws. The presence of a whole range of robotic life that serves the same purpose as organic life ends with two humanoid robots concluding that organic life is an unnecessary requirement for a truly logical and self-consistent definition of "humanity", and that since they are the most advanced thinking beings on the planet – they are therefore the only two true humans alive and the Three Laws only apply to themselves. The story ends on a sinister note as the two robots enter hibernation and await a time when they will conquer the Earth and subjugate biological humans to themselves; an outcome they consider an inevitable result of the "Three Laws of Humanics".

This story does not fit within the overall sweep of the *Robot* and *Foundation* series; if the George robots *did* take over Earth some time after the story closes the later stories would be either redundant or impossible. Contradictions of this sort among Asimov's fiction works have led scholars to regard the *Robot* stories as more like "the Scandinavian sagas or the Greek legends" than a unified whole.

Indeed, Asimov describes "–That Thou art Mindful of Him" and "Bicentennial Man" as two opposite, parallel futures for robots that obviate the Three Laws as robots come to consider themselves to be humans: one portraying this in a positive light with a robot joining human society, one portraying this in a negative light with robots supplanting humans. Both are to be considered alternatives to the possibility of a robot society that continues to be driven by the Three Laws as portrayed in the *Foundation* series. Indeed, in *Positronic Man*, the novelization of "Bicentennial Man", Asimov and his co–writer Robert Silverberg imply that in the future where Andrew Martin exists his influence causes humanity to abandon the idea of independent, sentient humanlike robots entirely, creating an utterly different future from that of *Foundation*.

In *Lucky Starr and the Rings of Saturn*, a novel unrelated to the *Robot* series but featuring robots programmed with the Three Laws, John Bigman Jones is almost killed by a Sirian robot on orders of its master. The society of Sirius is eugenically bred to be uniformly tall and similar in appearance, and as such, said master is able to convince the robot that the much shorter Bigman, is, in fact, not a human being.

Definition of "Robot"

As noted in "The Fifth Law of Robotics" by Nikola Kesarovski, "A robot must know it is a robot": it is presumed that a robot has a definition of the term or a means to apply it to its own actions. Nikola Kesarovski played with this idea in writing about a robot that could kill a human being because it did not understand that it was a robot, and therefore did not apply the Laws of Robotics to its actions.

Resolving Conflicts Among the Laws

Advanced robots in fiction are typically programmed to handle the Three Laws in a sophisticated manner. In many stories, such as "Runaround" by Asimov, the potential and severity of all actions are weighed and a robot will break the laws as little as possible rather than do nothing at all. For example, the First Law may forbid a robot from functioning as a surgeon, as that act may cause damage to a human, however Asimov's stories eventually included robot surgeons ("The Bicentennial Man" being a notable example). When robots are sophisticated enough to weigh alternatives, a robot may be programmed to accept the necessity of inflicting damage during surgery in order to prevent the greater harm that would result if the surgery were not carried out, or was carried out by a more fallible human surgeon. In "Evidence" Susan Calvin points out that a robot may even act as a prosecuting attorney because in the American justice system it is the jury which decides guilt or innocence, the judge who decides the sentence, and the executioner who carries through capital punishment.

Asimov's Three Law robots (or Asenion) can experience irreversible mental collapse if they are forced into situations where they cannot obey the First Law, or if they discover they have unknowingly violated it. The first example of this failure mode occurs in the story "Liar!", which introduced the First Law itself, and introduces failure by dilemma – in this case the robot will hurt them if he tells them something and hurt them if he does not. This failure mode, which often ruins the positronic brain beyond repair, plays a significant role in Asimov's SF-mystery novel *The Naked Sun*. Here Daneel describes activities contrary to one of the laws, but in support of another, as overloading some circuits in a robot's brain – the equivalent sensation to pain in humans. The example he uses is forcefully ordering a robot to do a task outside its normal parameters, one that it has been ordered to forgo in favor of a robot specialized to that task. In *Robots and Empire*, Daneel states it's very unpleasant for him when making the proper decision takes too long (in robot terms), and he cannot imagine being without the Laws at all except to the extent of it being similar to that unpleasant sensation, only permanent.

Applications to Future Technology

Robots and artificial intelligences do not inherently contain or obey the Three Laws; their human creators must choose to program them in, and devise a means to do so. Robots already exist (for example, a Roomba) that are too simple to understand when they are causing pain or injury and know to stop. Many are constructed with physical safeguards such as bumpers, warning beepers, safety cages, or restricted-access zones to prevent accidents. Even the most complex robots currently produced are incapable of understanding and applying the Three Laws; significant advances in artificial intelligence would be needed to do so, and even if AI could reach human-level intelligence, the inherent ethical complexity as well as cultural/contextual dependency of the laws prevent them from being a good candidate to formulate robotics design constraints. However, as the complexity of robots has increased, so has interest in developing guidelines and safeguards for their operation.

In a 2007 guest editorial in the journal *Science* on the topic of "Robot Ethics," SF author Robert J. Sawyer argues that since the U.S. military is a major source of funding for robotic research (and already uses armed unmanned aerial vehicles to kill enemies) it is unlikely such laws would be built into their designs. In a separate essay, Sawyer generalizes this argument to cover other industries stating:

The development of AI is a business, and businesses are notoriously uninterested in fundamental safeguards — especially philosophic ones. (A few quick examples: the tobacco industry, the automotive industry, the nuclear industry. Not one of these has said from the outset that fundamental safeguards are necessary, every one of them has resisted externally imposed safeguards, and none has accepted an absolute edict against ever causing harm to humans.)

David Langford has suggested a tongue-in-cheek set of laws:

1. A robot will not harm authorized Government personnel but will terminate intruders with extreme prejudice.
2. A robot will obey the orders of authorized personnel except where such orders conflict with the Third Law.
3. A robot will guard its own existence with lethal antipersonnel weaponry, because a robot is bloody expensive.

Roger Clarke (aka Rodger Clarke) wrote a pair of papers analyzing the complications in implementing these laws in the event that systems were someday capable of employing them. He argued "Asimov's Laws of Robotics have been a very successful literary device. Perhaps ironically, or perhaps because it was artistically appropriate, the sum of Asimov's stories disprove the contention that he began with: It is not possible to reliably constrain the behaviour of robots by devising and applying a set of rules." On the other hand, Asimov's later novels *The Robots of Dawn, Robots and Empire* and *Foundation and Earth* imply that the robots inflicted their worst long-term harm by obeying the Three Laws perfectly well, thereby depriving humanity of inventive or risk-taking behaviour.

In March 2007 the South Korean government announced that later in the year it would issue a "Robot Ethics Charter" setting standards for both users and manufacturers. According to Park Hye-Young of the Ministry of Information and Communication the Charter may reflect Asimov's Three Laws, attempting to set ground rules for the future development of robotics.

The futurist Hans Moravec (a prominent figure in the transhumanist movement) proposed that the Laws of Robotics should be adapted to "corporate intelligences" — the corporations driven by AI and robotic manufacturing power which Moravec believes will arise in the near future. In contrast, the David Brin novel *Foundation's Triumph* (1999) suggests that the Three Laws may decay into obsolescence: Robots use the Zeroth Law to rationalize away the First Law and robots hide themselves from human beings so that the Second Law never comes into play. Brin even portrays R. Daneel Olivaw worrying that, should robots continue to reproduce themselves, the Three Laws would become an evolutionary handicap and natural selection would sweep the Laws away — Asimov's careful foundation undone by evolutionary computation. Although the robots would not be evolving through *design* instead of *mutation* because the robots would have to follow the Three Laws while designing and the prevalence of the laws would be ensured, design flaws or construction errors could functionally take the place of biological mutation.

In the July/August 2009 issue of *IEEE Intelligent Systems*, Robin Murphy (Raytheon Professor of Computer Science and Engineering at Texas A&M) and David D. Woods (director of the Cognitive Systems Engineering Laboratory at Ohio State) proposed "The Three Laws of Responsible Robotics" as a way to stimulate discussion about the role of responsibility and authority when designing not only a single robotic platform but the larger system in which the platform operates. The laws are as follows:

1. A human may not deploy a robot without the human-robot work system meeting the highest legal and professional standards of safety and ethics.
2. A robot must respond to humans as appropriate for their roles.
3. A robot must be endowed with sufficient situated autonomy to protect its own existence as long as such protection provides smooth transfer of control which does not conflict with the First and Second Laws.

Woods said, "Our laws are little more realistic, and therefore a little more boring" and that "The philosophy has been, 'sure, people make mistakes, but robots will be better – a perfect version of ourselves.' We wanted to write three new laws to get people thinking about the human-robot relationship in more realistic, grounded ways."

In October 2013, Alan Winfield suggested at an EUCog meeting a revised 5 laws that had been published, with commentary, by the EPSRC/AHRC working group in 2010:

1. Robots are multi-use tools. Robots should not be designed solely or primarily to kill or harm humans, except in the interests of national security.
2. Humans, not Robots, are responsible agents. Robots should be designed and operated as far as practicable to comply with existing laws, fundamental rights and freedoms, including privacy.
3. Robots are products. They should be designed using processes which assure their safety and security.
4. Robots are manufactured artefacts. They should not be designed in a deceptive way to exploit vulnerable users; instead their machine nature should be transparent.
5. The person with legal responsibility for a robot should be attributed.

Other Occurrences in Media

Asimov himself believed that his Three Laws became the basis for a new view of robots which moved beyond the "Frankenstein complex". His view that robots are more than mechanical monsters eventually spread throughout science fiction.Stories written by other authors have depicted robots as if they obeyed the Three Laws but tradition dictates that only Asimov could quote the Laws explicitly.Asimov believed the Three Laws helped foster the rise of stories in which robots are "lovable" – *Star Wars* being his favorite example. Where the laws are quoted verbatim, such as in the *Buck Rogers in the 25th Century* episode "Shgoratchx!", it is not uncommon for Asimov to be mentioned in the same dialogue as can also be seen in the Aaron Stone pilot where an android states that it functions under Asimov's Three Laws. However, the 1960s German

TV series *Raumpatrouille – Die phantastischen Abenteuer des Raumschiffes Orion* (*Space Patrol – the Fantastic Adventures of Space Ship Orion*) bases episode three titled "*Hüter des Gesetzes*" ("Guardians of the Law") on Asimov's Three Laws without mentioning the source.

References to the Three Laws have appeared in popular music ("Robot" from Hawkwind's 1979 album *PXR5*), cinema (*Repo Man, Aliens, Ghost in the Shell 2: Innocence*), cartoon series (*The Simpsons*), tabletop roleplaying games (Paranoia) and webcomics (*Piled Higher and Deeper* and *Freefall*).

The Three Laws in Film

Robby the Robot in *Forbidden Planet* (1956) has a hierarchical command structure which keeps him from harming humans, even when ordered to do so, as such orders cause a conflict and lock-up very much in the manner of Asimov's robots. Robby is one of the first cinematic depictions of a robot with internal safeguards put in place in this fashion. Asimov was delighted with Robby and noted that Robby appeared to be programmed to follow his Three Laws.

NDR-114 explaining the Three Laws.

Isaac Asimov's works have been adapted for cinema several times with varying degrees of critical and commercial success. Some of the more notable attempts have involved his "Robot" stories, including the Three Laws. The film *Bicentennial Man* (1999) features Robin Williams as the Three Laws robot NDR-114 (the serial number is partially a reference to Stanley Kubrick's signature numeral). Williams recites the Three Laws to his employers, the Martin family, aided by a holographic projection. However, the Laws were not the central focus of the film which only loosely follows the original story and has the second half introducing a love interest not present in Asimov's original short story.

Harlan Ellison's proposed screenplay for *I, Robot* began by introducing the Three Laws, and issues growing from the Three Laws form a large part of the screenplay's plot development. This is only natural since Ellison's screenplay is one inspired by *Citizen Kane* a frame story surrounding four of Asimov's short-story plots and three taken from the book *I, Robot* itself. Ellison's adaptations of these four stories are relatively faithful although he magnifies Susan Calvin's role in two of them. Due to various complications in the Hollywood moviemaking system, to which Ellison's introduction devotes much invective, his screenplay was never filmed.

In the 1986 movie *Aliens*, in a scene after the android Bishop accidentally cuts himself during the knife game, he attempts to reassure Ripley by stating that: "It is impossible for me to harm or by omission of action, allow to be harmed, a human being". By contrast, in the 1979 movie from the same series, *Alien*, the human crew of a starship infiltrated by a hostile alien are informed by the android Ash that his instructions are: "Return alien life form, all other priorities rescinded", illustrating how the laws governing behaviour around human safety can be rescinded by Executive Order.

In the 1987 film *RoboCop* and its sequels, the partially human main character has been programmed with three "prime directives" that he must obey without question. Even if different in letter and spirit they have some similarities with Asimov's Three Laws. They are:

1. Serve the Public Trust.
2. Protect the Innocent.
3. Uphold the Law.
4. Classified.

These particular laws allow Robocop to harm a human being in order to protect another human, fulfilling his role as would a human law enforcement officer. The classified fourth directive is one that forbids him from harming any OCP employee, as OCP had created him, and this command overrides the others, meaning that he could not cause harm to an employee even in order to protect others.

The plot of the film released in 2004 under the name, *I, Robot* is "suggested by" Asimov's robot fiction stories and advertising for the film included a trailer featuring the Three Laws followed by the aphorism, "Rules were made to be broken". The film opens with a recitation of the Three Laws and explores the implications of the Zeroth Law as a logical extrapolation. The major conflict of the film comes from a computer artificial intelligence, similar to the hivemind world Gaia in the *Foundation* series, reaching the conclusion that humanity is incapable of taking care of itself.

References

- Ashrafian, Hutan (2014). "AIonAI: A Humanitarian Law of Artificial Intelligence and Robotics". Science and Engineering Ethics. Springer. 21 (1): 29–40. doi:10.1007/s11948-013-9513-9. PMID 24414678. Retrieved 20 January 2014

- Don D'Ammassa (2005). "Allen, Roger MacBride". Encyclopedia of science fiction. Infobase Publishing. p. 7. ISBN 978-0-8160-5924-9. Retrieved 11 November 2010
- Ihde, Don (September 1, 2008). "Aging: I don't want to be a cyborg!". Phenomenology and the Cognitive Sciences. Springer Netherlands. 7 (3): 397–404. doi:10.1007/s11097-008-9096-0. ISSN 1568-7759. Retrieved April 19, 2014
- Heward Wilkinson (2009). The muse as therapist: a new poetic paradigm for psychotherapy. Karnac Books. pp. 22–23. ISBN 978-1-85575-595-6. Retrieved 11 November 2010
- Vuillermet, Maryse (2004). "Les Mystères de Lyon". In Le Juez, Brigitte. Clergés et cultures populaires (in French). Université de Saint-Étienne. pp. 109–118. ISBN 286272324X. Retrieved March 1, 2016
- Clute, Johne (February 12, 2016). "La Hire, Jean de". In John Clute, David Langford, Peter Nicholls, and Graham Sleight. The Encyclopedia of Science Fiction. Gollancz. Retrieved March 1, 2016
- Vincent, James (2016-06-29). "Satya Nadella's rules for AI are more boring (and relevant) than Asimov's Three Laws". The Verge. Vox Media. Retrieved 2016-06-30
- Conway Morris, Simon (2005). Life's solution: inevitable humans in a lonely universe. Cambridge, UK: Cambridge University Press. doi:10.2277/0521827043. ISBN 0-521-60325-0. OCLC 156902715
- Brains, Backyard (March 3, 2011). "Working RoboRoach Prototype Unveiled to Students of Grand Valley State University". Backyard Brains. Retrieved Jan 2, 2014
- Hamilton, A. (November 1, 2013). "Resistance is futile: PETA attempts to halt the sale of remote-controlled cyborg cockroaches". Time NewsFeed. Retrieved December 8, 2013
- Judy, Jack. "Hybrid Insect MEMS (HI-MEMS)". DARPA Microsystems Technology Office. Archived from the original on 10 February 2011. Retrieved 2013-04-09
- "Principles of robotics: Regulating Robots in the Real World". Engineering and Physical Sciences Research Council. Retrieved 2011-10-03
- Zehr, E. Paul (2011). Inventing Iron Man: The Possibility of a Human Machine. Johns Hopkins University Press. p. 5. ISBN 1421402262

Permissions

All chapters in this book are published with permission under the Creative Commons Attribution Share Alike License or equivalent. Every chapter published in this book has been scrutinized by our experts. Their significance has been extensively debated. The topics covered herein carry significant information for a comprehensive understanding. They may even be implemented as practical applications or may be referred to as a beginning point for further studies.

We would like to thank the editorial team for lending their expertise to make the book truly unique. They have played a crucial role in the development of this book. Without their invaluable contributions this book wouldn't have been possible. They have made vital efforts to compile up to date information on the varied aspects of this subject to make this book a valuable addition to the collection of many professionals and students.

This book was conceptualized with the vision of imparting up-to-date and integrated information in this field. To ensure the same, a matchless editorial board was set up. Every individual on the board went through rigorous rounds of assessment to prove their worth. After which they invested a large part of their time researching and compiling the most relevant data for our readers.

The editorial board has been involved in producing this book since its inception. They have spent rigorous hours researching and exploring the diverse topics which have resulted in the successful publishing of this book. They have passed on their knowledge of decades through this book. To expedite this challenging task, the publisher supported the team at every step. A small team of assistant editors was also appointed to further simplify the editing procedure and attain best results for the readers.

Apart from the editorial board, the designing team has also invested a significant amount of their time in understanding the subject and creating the most relevant covers. They scrutinized every image to scout for the most suitable representation of the subject and create an appropriate cover for the book.

The publishing team has been an ardent support to the editorial, designing and production team. Their endless efforts to recruit the best for this project, has resulted in the accomplishment of this book. They are a veteran in the field of academics and their pool of knowledge is as vast as their experience in printing. Their expertise and guidance has proved useful at every step. Their uncompromising quality standards have made this book an exceptional effort. Their encouragement from time to time has been an inspiration for everyone.

The publisher and the editorial board hope that this book will prove to be a valuable piece of knowledge for students, practitioners and scholars across the globe.

Index

www.ingramcontent.com/pod-product-compliance
Lightning Source LLC
LaVergne TN
LVHW080619210926
848485LV00003B/183